THE CASE OF GALILEO

THE CASE OF
Galileo

A Closed Question?

ANNIBALE FANTOLI

Translated by George V. Coyne, S.J.

University of Notre Dame Press

Notre Dame, Indiana

Copyright © 2012 by University of Notre Dame
Notre Dame, Indiana 46556
www.undpress.nd.edu
All Rights Reserved

Manufactured in the United States of America

Library of Congress Cataloging-in-Publication Data

Fantoli, Annibale, 1924–
[Galileo e la Chiesa. English]
The case of Galileo : a closed question? /
Annibale Fantoli ; translated by George V. Coyne.
p.   cm.
Includes bibliographical references and index.
ISBN 978-0-268-02891-6 (pbk. : alk. paper) —
ISBN 0-268-02891-5 (pbk. : alk. paper)
E-ISBN: 978-0-268-07972-7
1. Galilei, Galileo, 1564–1642.   2. Galilei, Galileo, 1564–1642—
Trials, litigation, etc.   3. Catholic Church—History—17th century.
4. Religion and science—Italy—History—17th century.   I. Title.
QB36.G2F2613   2010
520.92—dc23

2012000866

♻ *This book is printed on recycled paper.*

For Marcello

CONTENTS

P R E S E N T A T I O N

The reason I feel especially privileged to present this work of Annibale Fantoli is found in the fact that it responds in my opinion in an exemplary way to the wish of John Paul II in his 1979 address on the occasion of the centenary of the birth of Albert Einstein, when he spoke of desiring to establish a commission to restudy the Galileo Affair:

> I hope that theologians, scholars, and historians, animated by a spirit of sincere collaboration, will study the Galileo case more deeply and in a loyal recognition of wrongs from whatsoever side they came, will dispel the mistrust that still opposes in many minds a fruitful concord between science and faith, between the Church and the world.

Fantoli is already well known among Galileo scholars for his extensive volume, originally in Italian: *Galileo: For Copernicanism and for the Church*. That book is already in its third edition in English and has been translated into Russian, Polish, French, Portuguese, Spanish, and Japanese. In fact, the revisions appearing in the third English edition of 2003, many of them quite significant, inspired Fantoli to prepare this new publication, which offers to the general educated public, who are not necessarily Galileo scholars, a clear updated synthesis of the many complex cultural factors that have shaped the history of the Galileo Affair. It is one of the best presentations available to the general public. Fantoli has the rare talent of combining a profound respect for the Church with an equally deep respect for historical truth; and he does that without assuming an apologetic posture.

This talent of his is particularly evident in the final part of this volume where he describes the history following the condemnation of Galileo up to our own times. It is the history of a Church that continues to bear the heavy burden of the Galileo Affair because of its constant preoccupation with saving its good name, while unwilling to accept, without shadowy compromise and veiled formulations, its own responsibility in the affair.

We have the good fortune to live in times when the dialogue among science, philosophy, and theology is heartily encouraged. Galileo did not have that good fortune. He had to battle from more than one trench. With the passage of time he has been proved correct. After having been defeated temporarily, he has triumphed, as history has subsequently confirmed, on his own merits on all fronts. It is a tribute to Fantoli to have shown clearly the special significance of this posthumous triumph of Galileo by dramatically showing that, in fact, it was precisely the lack of true dialogue that led to the tragic errors that caused such great suffering to Galileo and so much damage to the Church. Fantoli's work makes a significant contribution to this much sought after dialogue by teaching us that only humility and a sense of freedom can create in the human spirit the propensity to recognize the truth from whatever side it comes, an essential condition to avoiding future cases such as that of Galileo.

George V. Coyne, S.J., Director, Vatican Observatory
Castel Gandolfo, Rome, September 2005

After the publication of my work *Galileo: Per il Copernicanesimo e per la Chiesa,* published by the Vatican Observatory in 1993, friends suggested that I prepare a shorter version of it, aimed at a wider, nonspecialized public. In the following years, however, I was busy with the translations of the book into several languages, among which the English version, *Galileo: For Copernicanism and for the Church,* appeared in 1994, with a third edition in 2003. It was only towards the end of this period that I was able to prepare a shorter version, which appeared in Italian with the title *Il caso Galileo* (2003) and is now presented in its English translation.

This work follows in its main lines the text of the above-mentioned third English edition, except for the first part, which has been completely restructured. The very numerous and long notes have been eliminated, and the most important incorporated in the text in an abridged form. Further improvements have been introduced, taking into account the third Italian edition of my original book, which appeared in March 2010. All this will help, I hope, to give to the cultivated reader a better understanding of the complex philosophical, theological, and scientific factors that played such a decisive role in the origin and in the following development of the Copernican issue, with Galileo's drama at its center. Furthermore, it will show how difficult was the long road that brought the Catholic Church to a final admission (even though a very cautious and not fully satisfactory one) of its responsibility for the whole Galileo Affair.

I wish to express here all my gratitude to Father George V. Coyne, S.J., who while still director of the Vatican Observatory, wrote the Presentation of the book on the occasion of its first publication in Italian, and was subsequently able to find the time to translate this work into English, as he had already done on the occasion of all my previous works. I am also extremely grateful to Professor Wilbur Applebaum and to Naomi Polansky for their very generous, patient, and accurate review of the English text, and the many precious suggestions they have made concerning it. Last, but not least, my thanks go also to the University of Notre Dame Press, for the decision to publish this book, a decision that highly honors me.

Finally, let me add a brief note about citations in the text. I did not want to overburden readers of this volume with the scholarly apparatus present in my other, fully annotated volume on Galileo. Thus references in this volume use a short form providing basic author and title information; fuller information for those sources can be found in the bibliography. Except where noted otherwise in the citations, translations from documents contemporary to Galileo have been made by Father Coyne.

Victoria, British Columbia, December 2010

Prologue

It is June 22, 1633, in the morning. In a hall of the convent of Santa Maria sopra Minerva in Rome's historical center, an accused man, on his knees before seven cardinals and officials of the Congregation of the Holy Office as witnesses to the proceedings, listens to the decree of condemnation:

> We, Gasparo Borgia [et al.] . . . , by the grace of God, Cardinals of the Holy Roman Church, and especially commissioned by the Holy Apostolic See as Inquisitors-General against heretical depravity in all of Christendom . . . say, pronounce, sentence, and declare that you, the above-mentioned Galileo, because of the things deduced in the trial and confessed by you as above, have rendered yourself according to this Holy Office vehemently suspected of heresy, namely of having held and believed a doctrine which is false and contrary to the divine and Holy Scripture: that the sun is the center of the world and does not move from east to west, and the earth moves and is not the center of the world, and that one may hold and defend as probable an opinion after it has been declared and defined contrary to Holy Scripture. Consequently you have incurred all the censures and penalties imposed and promulgated by the sacred

1

canons and all particular and general laws against such delinquents. We are willing to absolve you from them provided that first, with a sincere heart and unfeigned faith, in front of us you abjure, curse, and detest the above-mentioned errors and heresies, and every other error and heresy contrary to the Catholic and Apostolic Church, in the manner and form we will prescribe to you.

Furthermore, so that this serious and pernicious error and transgression of yours does not remain completely unpunished, and so that you will be more cautious in the future and an example for others to abstain from similar crimes, we order that the book *Dialogue* by Galileo Galilei be prohibited by public edict.

We condemn you to formal imprisonment in this Holy Office at our pleasure. As a salutary penance we impose on you to recite the seven penitential Psalms once a week for the next three years. And we reserve the authority to moderate, change, or condone wholly or in part the above-mentioned penalties and penances. (Galileo, *Opere,* 19:402–6; trans. Finocchiaro, *Galileo Affair,* 289–91)

After the reading of the sentence, Galileo had no option but to obey. Still kneeling down he read the formula of abjuration that had been presented to him:

I, Galileo, son of the late Vincenzio Galilei of Florence, seventy years of age, arraigned personally for judgment, kneeling before you Most Eminent and Most Reverend Cardinals Inquisitors-General against heretical depravity in all of Christendom, having before my eyes and touching with my hands the Holy Gospels, swear that I have always believed, I believe now, and with God's help I will believe in the future all that the Holy Catholic and Apostolic Church holds, preaches, and teaches. However, whereas, after having been judicially instructed with injunction by the Holy Office to abandon completely the false opinion that the sun is the center of the world and does not move and the earth is not the center of the world and moves, and not to hold, defend, or teach this false doctrine in any way whatever, orally or in writing; and after having been notified that this doctrine is contrary to Holy Scripture; I wrote and published a book in which I treat of this already condemned doctrine and adduce very effective reasons in its favor, without refuting them in any way; therefore, I have been judged vehemently' suspected of heresy, namely of having held and believed that the sun is the center of the world and motionless and the earth is not the center and moves.

Therefore, desiring to remove from the minds of Your Eminences and every faithful Christian this vehement suspicion, rightly conceived against me, with a sincere heart and unfeigned faith I abjure, curse, and detest the above-mentioned errors and heresies, and in general each and every other error, heresy, and sect contrary to the Holy Church; and I swear that in the future I will never again say or assert, orally or in writing, anything which might cause a similar suspicion about me; on the contrary, if I should come to know any heretic or anyone suspected of heresy, I will denounce him to this Holy Office, or to the Inquisitor or Ordinary of the place where I happen to be.

Furthermore, I swear and promise to comply with and observe completely all the penances which have been or will be imposed upon me by this Holy Office; and should I fail to keep any of these promises and oaths, which God forbid, I submit myself to all the penalties and punishments imposed and promulgated by the sacred canons and other particular and general laws against similar delinquents. So help me God and these Holy Gospels of His, which I touch with my hands.

I, the above-mentioned Galileo Galilei, have abjured, sworn, promised, and obliged myself as above; and in witness of the truth I have signed with my own hand the present document of abjuration and have recited it word for word in Rome, at the convent of the Minerva, this twenty-second day of June 1633.

I, Galileo Galilei, have abjured as above, by my own hand. (Galileo, *Opere,* 19:406–7; trans. Finocchiaro, *Galileo Affair,* 292–93)

Thus one of the most famous trials in the history of Europe came to an end. At first, the condemnation of Galileo was known only among a small circle of educated people. But it would become, beginning especially in the age of the Enlightenment, emblematic of the inevitable opposition between the new worldview promoted by modern science and religious obscurantism, associated especially with the Catholic Church.

In this book I will try to follow from one stage to the next the growth in Galileo's manner of thinking and the consequent actions he took in favor of the new worldview presented in the *On the Revolutions of the Heavenly Spheres,* the epoch-making work of the great Polish astronomer Nicolaus Copernicus, a work that appeared a little over twenty years before Galileo's birth. I will also attempt to show how his activities brought the eminent scientist from Tuscany, despite his intentions, into conflict with a Catholic Church whose worldview was fixed in the old traditions, a conflict that would end in his condemnation by that very Church. Finally, I will follow the slow and painful process whereby the Church would come to accept the Copernican heliocentric worldview and finally admit in recent times to the errors it committed in condemning Galileo, an admission, however, quite guarded and in many ways unsatisfactory.

From Galileo's Birth
to His Teaching Years in Padua

Galileo was born in Pisa on February 15, 1564. At that time Italy was divided into many independent states, and Pisa, at one time a prosperous seafaring republic, was a part of the Grand Duchy of Tuscany, which was governed by the powerful Medici family, with its capital in Florence. At that time Florence was one of the richest cities in Europe. In the Middle Ages and especially in the Renaissance, Florence had made an incomparable contribution to Western art and culture. Galileo's family was from Florence. There had been a renowned medical doctor in the family whose name was also Galileo Galilei. It is quite probable that Galileo's father, Vincenzio, wished to give this name to his firstborn as a remembrance of his famous relative, and hoping that his son would follow in his footsteps in the medical profession. The family finances, at that time in a less than modest state, could thereby be put in order.

Vincenzio was a skillful lute player and an important member of the musical circle called the *Camerata Fiorentina,* where the theory of "drama in music" was developed. This eventually led to the Italian melodrama. But in order to make ends meet he was forced to engage in trading, and so, at Galileo's birth, the family was in Pisa.

As he grew up, Galileo gave clear signs of his extraordinary talents, and this only strengthened his father's plan to have him take up the profession of medicine. In September 1581 Galileo enrolled in the faculty of medicine at the University of Pisa. But to his father's great chagrin he discontinued his studies without having completed the course work. It was not so much that he was not content with the courses in medicine, which were still based on the writings of the famous Greek doctor, Galen (129–199 CE). His decision was rather attributed to his growing interest in the geometry of Euclid and the mechanics of Archimedes. Galileo had begun these studies under the tutelage of Ostilio Ricci (1540–1603), the mathematician who taught the pages of the grand duke of Tuscany. Galileo became fascinated by the rigor of mathematics together with experimentation in physics. He sensed that here lay his true vocation.

During the next four years Galileo deepened his knowledge of Euclid and especially of Archimedes. These studies prepared him for a brief period of teaching at Siena (1586–1587) and later at the University of Pisa, where in July 1589 he was appointed lecturer in mathematics. He had to teach, in addition to the geometry of Euclid, the two "classical" medieval treatises: the *Sphere* of Sacrobosco and the *Planetary Hypotheses*. Whether or not he had already come in contact with astronomy, this provided him the occasion to do so. And, as in all other European universities at that time, it was Ptolemaic astronomy that he had to study and teach at Pisa.

Ptolemy's astronomy (d. ca. 168 CE) came to be as an answer to many unresolved questions left by the theory of homocentric spheres that had been developed more than four centuries earlier by the Greek mathematician Eudoxus (409–365 BCE), who taught that the Earth is at the center of a complex system of spheres, the last of which had impressed upon it the so-called "fixed stars," almost all of the objects visible in the sky to the naked eye. In its daily axial rotation from east to west this sphere dragged along the seven planets that lay under it. But these seven planets all had quite irregular motions that varied from one to the other with stopping points and backward motions with respect to their west to east direct motions. Eudoxus had imagined these irregular motions as due to a combination of simple circular motions of one or more concentric spheres for each planet.

Aristotle (384–321 BCE) had adopted this system, and in his treatise *On the Heavens* had taken it as the foundation of the structure of the movements of all heavenly bodies. But the great Greek philosopher had also and above all else tried to fit the mathematical system of Eudoxus into a complete astrophysics.

According to Aristotle the physical makeup of the heavenly bodies is clearly different from that of the sublunary bodies that are centered on the Earth. The sublunary world is made up of four fundamental elements: earth, air, fire, and water. Objects in this realm are generally made up of combinations of the four. Each of these elements has its own natural place. The natural place of the element earth is at the center of the universe. This is surrounded by a spherical shell of water, which in turn is surrounded by spherical shells of air and of fire, the latter of which extends out to just under the Moon. Should an element leave its natural place, it would tend to return to that place by its own natural motion, which is rectilinear towards the center for the heavy elements of earth and water and composites made mostly of them, and away from the center for the light elements of air and fire and composites made mostly of them. Since there are many bodies consisting of these elements and their compounds, and since there is a great contrast among their natural motions and their other characteristics, the sublunary world is subject to continuous changes.

The heavenly bodies, as well as the spheres that carry them along, are composed of a single element called the ether or the "fifth essence," and their natural motion is circular, as shown by our everyday view of them. Since they are made up of a single element and are free of any contrasts in their circular motions, which go on indefinitely in the same direction, these bodies are immutable.

The Aristotelian universe is finite, bounded by the sphere of the fixed stars. And it exists from all eternity. Even if He did not create it, God is the ultimate source of its cosmic dynamism. All of the motions of finite and imperfect beings that populate the material world have, in fact, as the final cause the "desire" to be united with God, the perfect being and supreme good.

The fact that in this system the Earth is immobile at the center of the finite universe necessarily results from the "heaviness" of the element earth and of the composite bodies in which it is prevalent, and from its natural motion to go straight down to its natural place, the center of the universe. According to Aristotle this conclusion of "natural philosophy"—a term used until Newton's time to designate a branch of study similar to what we now call "science"—is confirmed by the experience of our senses, which do not detect any motion of the Earth, neither an axial rotation nor an orbital revolution about another body.

With this conviction Aristotle criticizes the theories of the Earth's motion developed by the school of Pythagoras. According to Philolaus (about 430 BCE) the Earth moves in the course of a day about a "central fire" (not to be confused with the Sun), which cannot be seen because the inhabited hemisphere (which contemporaries considered to be Europe, Africa, and Asia) of the Earth always faces away from it. That implied an axial rotation of the Earth. Later on the idea of a central fire was eliminated, but the axial rotation of the Earth remained. This idea was taken up by Heraclides Ponticus (388–310? BCE), a contemporary of Aristotle. These theories were labeled together as "Pythagorean," a term that at Galileo's time was used to denote also the more advanced theory of Aristarchus (see below), the only real forerunner of Copernicus's heliocentrism in Greek antiquity.

None of these theories, including that of Aristarchus, had the support of justification at the level of natural philosophy similar to that given by Aristotle in his cosmology, which on the contrary offered a truly grandiose vision of the world. So fascinating was Aristotle's system that it held sway in the teaching of natural philosophy in European universities right up to Galileo's time and even beyond. But from a strictly astronomical point of view Aristotle's system of homocentric spheres was readily seen as unsatisfactory. In fact, it required that each planet was at a constant distance from the Earth, and so it could not explain the increase and decrease of the apparent brightness of the planets with time, an explanation readily available if one posits changes in their distances from the Earth.

For a more satisfactory explanation of the heavenly motions it would be necessary to await the development of the mathematical concepts of eccentrics and epicycles by Apollonius (262–180 BCE) and by the great astronomer Hipparchus (d. 120 BCE). Relying upon those developments, Ptolemy, an astronomer and geographer from Alexandria, had written the greatest astronomical work of antiquity, the *Sintaxis,* which later became called the *Almagest,* meaning "the greatest," by Arabian astronomers.

As to physics, Ptolemy followed the cosmological view of Aristotle, namely, geocentrism. But instead of the theory of Eudoxus, which had been followed by Aristotle, Ptolemy had introduced an explanation of planetary motions founded on three principles: eccentric motions, epicycles, and the equant. By the first principle, the Earth was in a position slightly removed (eccentric) from the center of the planetary orbits. This easily explained both the variation during the year of the brightness of the planets and also the apparent variation in their velocities and thus, considering the Sun, that the seasons had unequal lengths. By the second principle, the motion of each planet results from a combination of more than one circular motion: each planet moves on a circle (epicycle) whose center is located on and moves along another larger circle (the deferent) which may itself rotate on another deferent and so on. The largest and final deferent is not centered on the Earth but on a point slightly displaced (eccentric) from the Earth. The result of this combination of circular motions is a trajectory called an epicycloid, which explained the systematic direct and retrograde motion of the planets. The third principle, the equant, is intended to explain the change in the angular velocity of the planets during the year. While every planet moves with uniform motion on its own epicycle, the center of the epicycle moves on the deferent with a constant angular velocity with respect to a point (equant) displaced from the center of the deferent by the same amount, but in the opposite direction, that the Earth is displaced.

Based on these three principles, Ptolemy finally succeeded in constructing tables (ephemerides) that gave the positions of the planets with time and that agreed reasonably well with observations. In particular, the use of epicycles proved to be quite pliant. By varying, as required, the radius of an epicycle and by adding on other epicycles one could correct previously computed positions so as to better fit the observations. Because of this pliancy, the Ptolemaic system remained for fourteen hundred years the alpha and omega of theoretical astronomy.

But there was a lingering fundamental question. Although the Ptolemaic system was undoubtedly satisfactory as a mathematical scheme, what physical significance did it have? Ptolemy himself was aware of the difficulty, and he tried to give physical meaning to his theory in the book *Planetary Hypotheses*. But his attempt, as well as the similar ones by medieval Arabian authors, did not convince the so-called natural philosophers, namely, those whom we might call the scientists of the time. Following Aristotle they claimed to know the physical structure of the world and that which caused the movements of both the heavenly and the Earthly bodies. And so there came to be a "divorce" between the views of the philosophers and those of the astronomers, who, in the wake of Ptolemy's thinking, continued to interest themselves in mathematical schemes that were useful for calculating celestial motions but were not very concerned about the physics behind those motions.

This situation lasted until the time of Galileo. Philosophy was considered superior to mathematics because it dealt with ultimate explanations, whereas mathematics was considered to be just a computing instrument. The difference became concretized in a higher academic status for philosophy in university teaching. The practical consequence was higher economic remuneration for philosophy teachers, including teachers of natural philosophy. And for a young reader in mathematics, such as Galileo, the salary was indeed meager.

Things being as they were, Galileo had to be careful not to push himself into the territory of his philosopher colleagues, but it was a situation that he was not prepared to endure forever. It was for him not just a question of prestige or economic well-being. He had a deep personal interest in enriching his knowledge of philosophy. He himself stated at a later date that he had "spent more years in studying philosophy than months in pure mathematics" (Galileo, *Opere,* 10:353). And certainly his key interest was in "natural philosophy."

The writings of Galileo during his time in Pisa bear witness to his interest in deepening his knowledge of philosophy. It is evident from those writings that the teaching of philosophy and astronomy by the Jesuits at the Roman College had an influence on him. The Society of Jesus was founded in 1540 by the Basque Ignatius of Loyola (1491–1556). Very soon after its founding it dedicated itself predominantly, but not exclusively, to teaching. And at Galileo's time the most active and well-known center of Jesuit teaching was that of the Roman College. Galileo had gone to Rome in the summer of 1587 and met there the famous Jesuit mathematician Christoph Clavius (1537–1612). They formed a friendship that would continue until Clavius's death. Galileo must have been deeply impressed by the academic level of the Jesuit instruction.

Among Galileo's writings during that period is the *Treatise on Heaven.* In this short treatise Galileo follows Aristotelian cosmology and makes it clear how much he depended on the texts used at the Roman College and, in particular, on Clavius's commentary on the *Sphere* of Sacrobosco, a medieval treatise on the astronomy of Ptolemy. And it is right from Clavius that Galileo derives reasons why the Copernican theory must be wrong (Galileo, *Opere,* 1:47–50). But before we examine what, in truth, was Galileo's personal position as regards Copernicanism, it would be useful to give a quick look at Copernicanism and at the history of its acceptance in the fifty years from Copernicus to Galileo's teaching in Pisa.

The great work of Copernicus (1473–1543), *On the Revolutions of the Heavenly Spheres* was published in 1543. Aristarchus of Samos (310–230 BCE) had already proposed in ancient Greece a Sun-centered system, so Copernicus's theory was not the *first* heliocentric system. But Copernicus gave a thorough mathematical treatment, so that his work was an absolutely new breakthrough in the world of science. The sweeping synthesis that it provided was such that the *On the Revolutions* could be compared to only one other such work, the *Almagest* of Ptolemy.

Copernicus claimed that to explain the phenomena in the heavens all that was required was to put the Sun, instead of the Earth, at the center and attribute three motions to the Earth: daily rotation on its own axis, which would explain the apparent daily motion of the heavenly bodies about the Earth; orbital motion about the Sun, which explains the apparent motion of the Sun along the ecliptic, the seasons, and the complex direct and retrograde motions of the planets; and the precession of the Earth's axis of rotation, which, according to Copernicus, was required to explain the constant inclination of 23.5 degrees of the Earth's rotation axis to the plane of the ecliptic.

These fundamental ideas, intended for nonspecialists, are found in book 1 of *On the Revolutions,* where Copernicus tried to answer the objections that Aristarchus's system had already had to face and that had hindered its success among the majority of the Greek philosophers of nature and astronomers. As far as science goes, the weightiest objection against the Earth's movement was undoubtedly the one based upon the absence of an observed parallax for the "fixed stars." They should, if the Earth is truly moving, be seen in different positions in the sky during the course of the year. And this effect had never been observed. Copernicus repeated the response of Aristarchus: the dimensions of the sphere of the "fixed stars" were, in comparison to the size of the Earth and the distance of the Earth from the Sun, so much larger that, although the parallax was there, it was too small to be measured.

Although this heliocentric system presented a major and basic change in the way of thinking of the celestial motions, in many aspects it remained faithful to long-standing traditions. For instance, the idea that the universe had a spherical shape and that all celestial motions were circular, which had never been challenged in two thousand years by western philosophers and astronomers, still held. But this made inevitable the introduction of a series of epicycles and eccentrics so that theory would match observations. And so the great advantage of the simplicity of the Copernican theory over that of Ptolemy, as Copernicus himself emphasized in Book I, was lost for the most part in the later mathematical developments of *On the Revolutions*. On the other hand, because he had to introduce eccentric motions with respect to the Sun, Copernicus's theory was strictly speaking no longer heliocentric.

To conclude, the Copernican system qualitatively had the undeniable advantage of simplicity. And it undoubtedly was better than the Ptolemaic system in the mathematical description of the motions of the inferior planets, Mercury and Venus, as well as that of the Moon. The result was that many of the most renowned astronomers of the time preferred Copernicanism as a mathematical theory to calculate planetary motions. But they refused to accept it as a physical explanation of how the world really worked. In fact, it apparently contradicted sense experience and the principles of Aristotelian cosmology. Of greater importance still it appeared to be in clear contradiction with scriptural statements about the stability of the Earth and the motion of the Sun.

Copernicus himself felt the weight of these difficulties and feared critical reactions from the Aristotelian camp, where the teaching of natural philosophy in European universities was still concentrated, as well as from theologians. Without a doubt, this was the main reason for his hesitation in publishing his work. In fact, however, the reactions he dreaded did not materialize, at least not with the virulence he feared. This fair-weather situation could be attributed in part to the influence of the "Note to the Reader," an anonymous preface to the book by the Protestant editor Andreas Osiander (1498–1552), who wrote that Copernicus's theory should be considered to be a pure mathematical hypothesis and not a physical explanation of the universe. Since it was only a mathematical theory, Osiander added, it was not more probable than any of the old theories. Still, it was published because of the "admirable hypotheses" it contained and, above all, because of its simplicity. "But as concerns hypotheses," Osiander repeated, "no one expects any kind of certainty from astronomy, which is not capable of providing such certainties." As we see, this harks back to the old thesis of the complete divorce between the areas of competence of natural philosophy and astronomical theories.

Copernicus was spared the sad experience of witnessing this betrayal of his real intentions. Having already suffered a cerebral paralysis some months before, he was dying when a copy of *On the Revolutions* was put in his hands. About five hundred copies of the book were printed. The second edition appeared only after twenty-three years and the third only after another fifty-one years. It would be without a doubt a mistake to state that the work of Copernicus received no attention from astronomers and the educated class in Europe. But, for reasons already mentioned, the theory appeared to be unacceptable as a real explanation of the structure of the universe. So the number of those who adhered to the new view of the world remained for the moment quite small. In Germany the principal promoter was Michael Mästlin (1550–1631), who probably deserves the honor of having introduced the ideas of Copernicus to his great disciple, Johannes Kepler (1571–1630). Even earlier than in Germany there were in England those who sympathized with Copernicanism. Among these were Robert Recorde (1510–1558), the greatest English mathematician of that period, and the famous William Gilbert (1544–1603), author of the treatise *On Magnetism* (1600). Even more clearly in favor of Copernicanism was Thomas Digges (ca. 1546–1595) in his work *Wings or Mathematical Stairs* (1572) and then in his appendix to the work on meteorology of his father Leonard, *Prognostication Everlasting* (1572). To this list of more or less declared supporters of Copernicus one should add the Frenchman Pierre de la Ramée (1515–1572) and the Italian Giovanni Battista Benedetti (1530–1590) and, of course, the famous Giordano Bruno, of whom I will write shortly.

The clearest proof that heliocentrism did not succeed as a physical explanation among astronomers during the period immediately following Copernicus is found in the position espoused by the most famous of the period's astronomers, Tycho Brahe (1546–1601). This Danish astronomer's principal claim to fame is his realization that a large quantity of observations, much more precise than had been achieved previously, would be required in order to construct a satisfactory theory of planetary motions. And so he dedicated many years to the ambitious project of acquiring such observations at Uraniborg, a center for astronomical observations without rival in Europe, which he had constructed on the island of Hven. The data thus accumulated by Brahe would be used later on by Kepler to discover his three laws of planetary motion, a discovery that would contribute decisively to overcoming the traditional view of the world.

But Brahe's claim to fame goes beyond that valuable collection of observations. In 1572 a nova, a new star, appeared in the constellation of Cassiopeia, and it immediately became a topic of lively discussion among astronomers as well as among Aristotelian philosophers. Was the nova a phenomenon in the Earth's atmosphere or did it belong to the heavenly spheres? The Aristotelian principle of the incorruptibility of the heavens required that it belong to the atmosphere, since the birth of a new star in the heavens would contradict that principle. But Brahe's accurate observations did not detect any parallax for the nova. He, therefore, deduced that it must be a distant object, at least beyond the Moon and more probably at the distance of or very close to the sphere of the fixed stars. His results were published in a very limited edition with the title *On the New Star* (1573). It was distributed only to a very limited circle of his correspondents and, therefore, did not exert much of an influence on the debate. Indeed, in the meantime the Jesuit mathematician Clavius had also arrived, most likely independently, at the same conclusions as Brahe, and he had published them in the 1585 edition of his

Commentary on the Sphere of Sacrobosco. But, like Brahe, Clavius was not able to give a plausible explanation for the nova as belonging to the heavenly world. And so, for the time being, the Aristotelian philosophers did not have to pay attention to such conclusions.

A new and even harder blow was given to the Aristotelian view by the appearance of numerous comets between 1577 and 1596. Given his authority gained by this time through his observations, it became ever more difficult to question Brahe's results. Proofs were being established through his observations that the comets belonged to the heavens and not to the sublunary regions. Thus, these new phenomena were giving lie to the dogma of the immutability of the heavenly bodies. Furthermore, the probability that the comets were moving about the Sun on noncircular orbits implied that they were able to cross unhindered the numerous solid spheres postulated by Aristotelian cosmology. The most obvious conclusion was to deny that these spheres existed at all. The crisis became even more acute with the posthumous publication of Brahe's work *Progymnasmata* (1602), which had a much larger distribution than the *On the New Star.* The authority of Brahe continued to make it ever more difficult to deny the observations of new phenomena in the heavens, an obvious contradiction of the Aristotelian dogma of their immutability.

Still Brahe could not accept the Copernican theory, even though he was one of the major contributors in bringing doubt upon the fundamental concepts of Aristotelian cosmology. Without a doubt the traditional difficulties stood in Brahe's way. Common experience argued against the movement of the Earth, for instance, in the absence of the strong winds that would be expected if the Earth moved. And then there were the objections from sacred scripture, which I address later. But the strongest objection arose from the fact that there was no measurable parallax of the fixed stars. Brahe certainly knew of Aristarchus's response, taken up by Copernicus, that it was the enormous distances of the stars that would not permit a measurement of the parallax that actually did exist. But Brahe had serious difficulties in accepting the postulated immensity of space. This difficulty arises from the fact that at the time of Brahe the apparent diameter of a star as seen by the naked eye was taken as a measure of its physical diameter. This was before the use of the telescope for astronomical observations and before any knowledge of light diffraction, which causes the apparent size of a star to be larger the brighter the star. So, if one admitted that the stars were at very large distances, then they must have sizes enormously greater than the Sun; of course, Brahe judged this to be very unlikely. Another difficulty for Brahe arose from the lack of any observations that the comets had both retrograde and direct motions in their orbits about the Sun.

With all of these objections in mind, the Danish astronomer devised his own theory of planetary motions. It incorporated the simplifying elements in the Copernican theory but maintained the Earth immobile at the center of the universe. In their daily motions, according to Brahe, the fixed stars and the planets revolve about the Earth, just as in the Ptolemaic system. The annual motion of the planets is explained in the following way: while the Moon and the Sun revolve around the Earth, the other five planets revolve about the Sun. Since Brahe embraced the traditional notion of circular motions, he had to employ epicycles and eccentrics to be able to fit the observations. Also, unlike Copernicus, he had to use the equants of Ptolemy.

From a mathematical point of view, Brahe's system was almost exactly the same as that of Copernicus. But by leaving the Earth at the universe's center, it avoided all of the objections to the Copernican theory and so met a certain favor among astronomers of his time, at least among those who kept to the area of mathematical hypotheses without worrying about physical reality. But for those who did not accept the complete divorce between mathematical hypotheses and natural philosophy, Brahe's system was still problematic. And, as we shall see, it is precisely because of considerations of physics that Galileo will refuse to consider the system of Brahe as a real alternative to the "two great world systems," the one of Aristotle and Ptolemy, the other of Copernicus.

It is totally improbable that Galileo had known at Pisa of the existence of Tycho Brahe's system, which was described for the first time in Brahe's book on the 1577 comet. The book was published in 1587. But, as we have seen, he certainly knew and probably read Copernicus's *On the Revolutions* while he was at Pisa. In another work of Galileo from that time, *On Motion,* he put forward the hypothesis of a "nonviolent," "neutral" rotation of a sphere at the center of the world. This appears to be one of his first attempts to consider that the Earth rotated. Does this mean that from that time on Galileo was leaning towards Copernicanism?

Such an interpretation is supported by Galileo's own statement that he made in a letter to Kepler in 1597. As we shall see, he states there that he had "already come many years ago to the opinion of Copernicus." Many Galilean scholars treat this statement with suspicion and prefer to reduce the "many years" to only a few years. I think, however, that we cannot rule out a literal interpretation of Galileo's words, not in the sense of a complete conversion to Copernicanism, which will not happen until 1610, but in the sense that he preferred it as a hypothesis that, as far back as his stay in Pisa, he saw as much more satisfactory than that of Aristotle and Ptolemy. It is possible that the heliocentric theory had made a deep impression on him at that time and that he was intuitively persuaded of its superiority over the traditional view. And perhaps he had begun to ponder physical proofs for Copernicanism. As we have seen, in his *On the Heavens* he had repeated the reasons why the Copernican hypothesis was to be rejected; but he had done this by quoting almost literally the words of Clavius's *Commentary of the Sphere of Sacrobosco* without necessarily making them his own. Still working within the framework of Aristotle's theory of natural places and natural motions, it could be that Galileo thought that he had found with his notion of nonviolent, neutral circular motion a physical justification for the Earth's rotation. Of course, this would have been only a first step towards Copernicanism, because that notion of circular motion was no longer applicable when it came to the orbital motion of the Earth about the Sun. He would, therefore, have had to sense, again intuitively, that a new theory of motion completely detached from that of Aristotle was required. This was undoubtedly a further reason for those studies on motion that he would carry forward during his time in Padua. For now, it is true, Galileo on this point was still in the dark. And he preferred to leave it there for the moment in expectation of a clearer view on it in the future.

From evidence gathered from Galileo's long stay at Padua (1592–1610) his growing adherence to heliocentrism becomes ever more explicit. Near the end of his three-year period of teaching at Pisa, Galileo sought out and obtained a position as mathematics teacher at the University of Padua, at that time one of the most prestigious universities of Europe. He was twenty-eight and would spend eighteen years at Padua, the best years of his life, as he would later write (Galileo, *Opere*, 18:209). It was certainly a very important time for the development of his studies on motion and especially that of falling bodies. These studies would lead him to a complete divorce from the fundamentals of the physics of Aristotle and to the formulation of some principles upon which modern physics is based. I would like now to trace his steps towards Copernicanism during this period.

As he had done at Pisa, Galileo also taught mathematics (Euclid's *Elements*) and astronomy at Padua. In his public and private teaching he still followed, as was expected of him, the astronomy of Ptolemy as contained in the *Sphere* of Sacrobosco, *Planetary Theory*, and *Questions in Mechanics*, attributed at that time to Aristotle. Since medical students also attended his lectures, they must have been given at a quite modest level of scholarship. In those times every respectable medical doctor was expected to be able to prepare horoscopes for his patients and so the lessons in astronomy were designed to provide the basic notions of astrology for their preparation.

Galileo's own work, *Treatise on the Sphere, or Cosmography*, of which there are various extant copies written by his disciples, is proof that he kept his teaching within the boundaries of traditional astronomy. But this in no way can be taken to mean that he was not, even then, directed towards Copernicanism, although Galileo was well aware that he had no proofs for it. Lacking such proofs he prudently preferred not to take positions. The natural philosophy professors, such as the famous Cesare Cremonini (1550?–1631), were very influential at Padua. To defend a thesis such as Copernicanism, diametrically opposed to that of Aristotle, especially without having proofs and a new natural philosophy that could ultimately justify such a thesis, would have meant exposing oneself to ridicule. And as a young professor Galileo absolutely had to avoid such circumstances.

Indeed, a preliminary idea of a possible proof of Copernicanism had probably come to him in about 1595 with his attempt to explain the tides—which are much more evident at the north end of the Adriatic Sea, especially at Venice, than elsewhere in the Mediterranean basin—as due to the two-fold rotary motion of the Earth required by the Copernican theory. Galileo would continue to develop this idea and would present it more than thirty years later in the *Dialogue* as one of his main arguments for Copernicanism. He argued that in the space of twenty-four hours the rotation of the Earth on its axis reinforced that of its revolution about the Sun during the first twelve hours and opposed it during the next twelve hours. These velocity variations created the movements of the sea masses, the tides, just as Galileo had seen happening in the water tankers that carried fresh water from Chioggia to Venice. The accelerations and the braking of the tankers caused the water to rise respectively towards the stern and towards the prow.

This sudden "intuition" must have pushed Galileo more and more to Copernicanism. But by temperament he was not inclined to sudden conversions of mind. On the contrary, he was surely well aware of the serious work required to come to a "proof" based on that intuition. In fact, his full and definitive adherence to heliocentrism will only come much later, and it would not be the tides that convinced him but his telescopic discoveries.

Two letters written by Galileo in 1597 confirm that by that time he considered the Copernican system to be much more probable than that of Aristotle and Ptolemy. The first was sent in May of that year to his old colleague at Pisa, Jacopo Mazzoni, who had written a book, *On the Complete Philosophy of Plato and Aristotle,* which included an argument against Copernicanism. Galileo judged the argument to be erroneous and he frankly confessed to his friend:

> But to tell the truth . . . I was confused and taken back by seeing Your Most Renowned Excellency so clearly resolute in impugning the opinion of the Pythagoreans and of Copernicus on the motion and location of the Earth. Since I have held that opinion to be much more probable than that of Aristotle and Ptolemy, I have made every effort to give a hearing to the reason offered by Your Excellency as I have an idea about this opinion and about all the matters linked to it. (Galileo, *Opere,* 2:198)

As we see, Galileo states that he holds the opinion of Copernicus to be "much more probable" than that of Aristotle and Ptolemy. And he adds that he has "an idea" not yet completely clear so that it remains an inkling about that opinion and about all the matters linked to it. It is quite probable that Galileo is alluding here to his theory of the tides. Galileo could surely have softened the way he expressed his support for Copernicanism so as not to injure his friend. But he uses words that appear to express well the degree to which by that time he was persuaded of the truth of Copernicanism.

Galileo's profession of Copernicanism in the second letter sent in August of that same year to Kepler appears to be even more explicit. The previous year Kepler had published his *Mystery of the Cosmos,* and two copies of it had been carried to Italy by his friend, Paul Hamberger, and delivered to Galileo. Apparently Hamberger took the initiative to do this, because at that time Kepler had not yet heard about the professor of mathematics at Padua. At any rate, Galileo wished to send a letter of thanks in Latin to Kepler. In it he admitted that he had read only the preface to Kepler's book, but he added that that reading was enough for him to become acquainted with Kepler's ideas. And he declared that he is happy to have in Kepler "such a companion in the search for truth." After a promise to read the book where "he was certain to find most wonderful things," he concluded:

And I will do it [read the book] even more eagerly in so far as I came to embrace the opinion of Copernicus already many years ago, and I have found in this hypothesis the explanation of many of nature's phenomena, which for sure remain unexplained by the current hypotheses. I have worked out many proofs and responses to arguments from the opposing camp, but I have not yet dared to make them public because I am frightened by what happened to Copernicus, our teacher, who, although from some he won immortal fame, was ridiculed and rejected by an infinite number of others (the stupid are so many). I would certainly not hesitate to put forward my thoughts, were there more persons like yourself. But since that is not the case, I will hold back. (Galileo, *Opere,* 10:68)

In this letter Galileo appears to distinguish two stages: that "already many years ago" he had come to the "opinion of Copernicus," and that "many of nature's phenomena" could be explained by it. This second stage certainly includes his "proof" from the tides, a fact that Kepler himself very perspicaciously intuited. But his idea of the tides only went back a few years. Did Galileo want to exaggerate the number of years so as not to appear less wise in Kepler's view? Many Galilean scholars think so. But we have already seen possible signs that during his years in Pisa he was being drawn towards Copernicanism. So, if he was exaggerating to Kepler, it remains to be proven. What is noteworthy is the cautious way in which Galileo expresses himself in this letter as well as in the previous one to Mazzoni. In fact, he speaks of the "opinion of Copernicus" and of "hypothesis." Even if he was by now persuaded that Copernicanism was much more probable than the theory of Aristotle and Ptolemy, he was aware that the elements in its favor were only at the point of departure. There was a great deal of work to be done. In the first place, it would be necessary to construct a new "philosophy of nature" to supplant that of Aristotle, specifically, to create a new theory of motion detached from the Aristotelian notion of natural places and natural motions, which, even as a pure theory, was an obstacle to any acceptance of Copernicanism. Right from his time in Pisa he began to work in that sense, and he would continue to do so during his years in Padua with quite significant results. As to the idea of the tides, he knew quite well that he would have to deepen his reasoning so that it could become a true scientific argument for the Earth's motion.

This attitude of Galileo, publicly teaching Ptolemaic astronomy on the one hand and on the other in his own mind taking up Copernicanism, has often been severely criticized, especially by authors who are not well disposed towards him, such as Arthur Koestler (in his book *The Sleeepwalkers*). But, in addition to the fact that they do not take account of the practicalities that dictated prudence, such criticisms seem to ignore that deeper meaning of Galileo's silence about his position, which is above all a proof of the seriousness of his scientific endeavors. By temperament Galileo was far from jumping at new ideas without reflecting on how to justify them, although he has been quite often unjustly accused of doing just that. Even after he intuited the truth of a theory, his scientific rigor would not allow him to rush to conclusions. Later on we will have the opportunity to see a similar reticence in the face of friends who would push him to take up a position more quickly.

Kepler responded to Galileo's letter two months later in October 1597. He was happy, he wrote, that their correspondence had begun and encouraged him to proceed with courage and even offered to have his writings on Copernicanism published in Germany, should he have found it difficult to do so in Italy. He also suggested to him that he make accurate measurements of two fixed stars that he identified in hopes of discovering parallax.

Galileo never replied to Kepler's letter. He probably never went much beyond reading the introduction to Kepler's *Mystery of the Cosmos*. It was difficult to read, and the thinking was permeated with notions that had little to do with science, at least as Galileo saw it. This must have made Galileo in a certain way uneasy with the German astronomer, and that feeling would last the rest of his life. Even when they began to correspond again after thirteen years it would almost always be Kepler, rather than Galileo, who showed a positive interest in pursuing their relationship.

Another letter to which Galileo never responded was one sent to him by Tycho Brahe in May of 1600. From a previous letter of Brahe to Vincenzo Pinelli, a friend of Galileo's from Padua (Galileo, *Opere*, 10:78–79), we know that the year before a disciple and future son-in-law of Brahe, Francesco Tegnagel, had visited Galileo and was told by him that he had read Brahe's *Astronomical Letters* (1596). Galileo also indicated that he intended to write to Brahe, but he never did so. Brahe, therefore, decided to write on his own. He spoke in the letter of his system for the planets and pointed out the advantages of it over that of Copernicus, inviting Galileo to discuss any points that he found interesting.

Why did Galileo never answer this letter from the most famous astronomer of that epoch? Possibly Galileo's statements to Tegnagel were simply a gesture of courtesy without any real wish on his part to start a correspondence with Brahe. As we shall see, throughout his life Galileo always kept an instinctive aversion for Brahe's system, which must, even from those early years, have appeared to him as a compromise without any real physical meaning. This aversion eventually led him to forget, or even to underestimate, the great merits of the Danish astronomer.

It was only in 1604 that Galileo first publicly expressed a criticism of Aristotelian cosmology when a nova was observed throughout Europe. Like the previous one in 1572, it led to much discussion among Aristotelian astronomers and philosophers. Yet again it was the Aristotelian doctrine of the immutability of the heavens that came up for discussion. The nova was seen at Padua for the first time on October 10, but Galileo did not observe it personally until October 28 (Galileo, *Opere*, 2:279). Since it aroused such great public interest, Galileo could not refrain from treating it. He probably did so with three lectures, of which there remain only the beginning and a fragment at the end. However, we also have some of Galileo's notes during that period, together with a collection he made of opinions on the nova of 1572, especially those of Tycho Brahe. Furthermore, the content of the lectures can be garnered from references to them in books published at that time and from one of Galileo's letters to be discussed shortly.

Using his own observations and those of his correspondents in other cities of Italy, Galileo had, in effect, declared that no parallax effect had been found for the nova during the period of observations and so he concluded that it must be "far beyond the orbit of the Moon" (Galileo, *Opere*, 10:134). He made no reference to Copernicanism since there was no direct connection between it and the interpretation of the nova. But from that time on he must have had some idea of using the observations of the nova as a possible argument in favor of Copernicanism. An indication of this is given by the fact that in his lectures he referred to the Latin philosopher Seneca, who speaks of the opinion of the Pythagoreans and of Aristarchus on the motion of the Earth.

Galileo's public lectures probably contributed to stirring up the discussions about the nova between the opposing camps. Among those opposed to Galileo, Cremonini was almost certainly the inspiration behind, if not the author of, the *Discourse on the New Star*, published in Padua in 1605 under the name of Antonio Lorenzini. According to "Lorenzini" the argument from the lack of parallax was founded upon sense experience and mathematics, applicable only to the realm of the Earth, and could not be applied to the heavenly spheres. In fact, it contradicted the unquestionable principles of the Aristotelian philosophy, which established the difference between the two worlds.

Probably aware of the true authorship of the *Discourse,* Galileo set himself to play the game and on his part inspired the composition of a *Dialogue* in the dialect of Padua. It was edited by a friend of his, the Benedictine Girolamo Spinelli, and was printed at Padua under the name of Cecco di Ronchitti (Galileo, *Opere,* 2:309–34) six weeks after the *Discourse.* The *Dialogue* questioned the Aristotelian distinction between the terrestrial and heavenly realms as well as the assertion that the heavenly bodies were incorruptible and that the Earthly bodies could not have a natural circular motion. On this latter matter, there was a quotation of "the lettered persons who say that the Earth turns round and round like a millstone" (2:322). By implication the assertion of the Earth's rotation was a denial of the rotation of the sphere of the fixed stars. Here also the *Dialogue* quoted that "there are quite a few (and good persons too) who believe that it [the sphere of the fixed stars] does not move" (2:318). Then the *Dialogue* claimed that the use of parallax for the fixed stars was justified and distinguished three kinds of parallax, a distinction that appeared to imply a Copernican interpretation of the nova.

But in what way did Galileo see this relationship between parallax and Copernicanism? He had noted a continuous decrease in the brightness of the nova. On the other hand, it seemed that the nova had formed in the region of Mars and Jupiter where it was first discovered. The only way to explain the lack of parallax was that the nova was moving away along the line of sight. But, if Copernicanism was correct, then observations made six months later should reveal a measurable parallax. Such a measurement would prove the motion of the Earth and, therefore, support Copernicanism.

This idea appears to be confirmed by the sketch of a letter from the following January in which Galileo asked pardon of his correspondent (unidentified) for not having sent a copy of his three public lectures. He wrote:

But since I, like many others, have the intention of placing before the judgment of the world my ideas not only about the location and the motion of this light [the nova] but also about its substance and origin, and since I believe that I have come upon a theory which has no obvious contradictions and so may be true, I find it necessary to be cautious and to go slowly by awaiting the return of this star in the east after it moves away from the Sun. Then very diligently I will observe what changes may have occurred both as to its location and as to the magnitude and quality of its brightness. As I continue my speculations about this marvelous star I have come to believe that there is more to be known than simple conjectures would suggest. And because this fantasy of mine draws out, or rather it puts forward, extremely important consequences and conclusions [from the planned observations of the nova] I have decided to change the text in a part of the discourse which I am composing about this topic. (Galileo, *Opere,* 10:134–35)

There is no doubt, in my opinion, that the words "extremely important consequences and conclusions" allude to an argument in favor of Copernicanism.

Why in the world was this letter never sent, and why was the *Discourse* never finished? It is likely that in the meantime Galileo began to have doubts which became stronger as time passed because a parallax was not observed even though months had gone by. Thus, the "proof" of Copernicanism he hoped to find eluded him, dashing his hopes to gain world renown. But this need not have implied that he abandoned his conviction of Copernicanism. The absence of parallax could be explained, without contradicting Copernicanism, by the fact that right from the beginning the nova had been located "high above all of the planets." This would only imply abandoning the idea, which Galileo had accepted, that the nova originated in the region of Mars and Jupiter. Given this immense "height" of the nova, no parallax could have been measured for it. So, Copernicanism was still a possibility, and for Galileo the most likely hypothesis. Lacking for the present an astronomical proof, there remained the physical one of the tides to work out. Galileo would have to wait, and he knew how to do that, as he will show in many other events of his life.

While waiting for a direct proof of Copernicanism, he must have sensed how important it was for him to proceed with his studies of motion, begun at Pisa. He would not have seen such studies as a "diversion" from his pursuit of proofs for Copernicanism nor even less so would they have meant that he was losing interest in finding such proofs. On the contrary, he must have been ever more convinced that it was necessary to set up a new science of motion and even more so a new "philosophy of nature" in general to serve as a physical justification for a heliocentric universe replacing that of Aristotle. During his years in Padua Galileo had already realized much progress in his studies in "physics," with fundamental results for constructing modern kinematics and dynamics. A systematic presentation of these results was given with the publication of his *Discourses* in 1638, for instance, his treatment both of the isochronic oscillations of the pendulum and of falling bodies.

While he was carrying forward these long-term researches, a completely fortuitous event occurred that had the potential to reopen the way to that "astronomical proof" for Copernicanism that had escaped him in 1604–5. A new instrument for observing appeared on the scene: the telescope. As is known, Galileo did not invent the telescope. It appears that the first rudimentary telescopes were made in Italy and in England beginning in the second half of the sixteenth century. Rather imperfect instruments were being sold in Flanders and in the Netherlands at the beginning of the seventeenth century. The Dutchman Hans Lippershey tried to obtain a patent for one in October 1608, and news of these happenings circulated in Europe. The Servite priest Paolo Sarpi, a friend of Galileo in Venice, received such a piece of news in November of that same year. A little later this was confirmed by an old student of Galileo at Padua, Jacques Badouère. Galileo possibly received information about the telescope from Sarpi himself on a visit Galileo made to Venice towards the end of July 1609. At about the same time a stranger had carried one of these instruments to Padua and then to Venice hoping to sell it to the Venetian Republic. According to his later statements Galileo did not see this instrument, but on the basis of written information he began right away to construct one in his workshop in Padua; at the end of August he had one ready to carry to Venice, and it was far superior to the one that the stranger was trying to sell. After he had given a practical demonstration to the dignitaries of the Republic of Venice, he received in official recognition of his "invention" an appointment as professor for life and a salary increase from 520 to 1000 florins a year.

Galileo was always being pressed economically, and so his construction of the telescope was probably motivated mainly by the need for money. Upon the death of his father in 1591, he, as the firstborn, had to assume the financial burden of maintaining the family in Florence. The provision of marriage dowries for his two sisters, Virginia and Livia, was particularly burdensome. In addition there were the expenses to maintain the family that he had borne at Padua from a Venetian woman, Marina Gamba. By her he had three children: Virginia in 1600, Livia in 1601, and Vincenzio in 1606. In order to keep up appearances, since he was not married to Marina, he had the added expense of keeping two separate dwellings. But beyond this commercial interest in the telescope Galileo realized very quickly its value as a scientific tool, and he began astronomical observations with it. Even in this Galileo cannot claim to be the first. Observations were done by others at least a year before. In 1609 Thomas Harriot, an Englishman, had tried to draft a map of the Moon by using the telescope. But to Galileo goes the credit for having fully realized, with the intuition of a true "philosopher of nature," the enormous importance of what he was observing.

His first discoveries during the autumn of 1609 encouraged him to make ever improved telescopes. With one that magnified twenty times, he succeeded beginning in December to obtain more accurate observations of the Moon and to discover, in January 1610, the satellites of Jupiter. Recognizing the truly revolutionary nature of his observations, especially those of the system of Jupiter, he could no longer hesitate to spread the word to the educated peoples of Europe. And so he published the *Starry Messenger* (Galileo, *Opere*, 3.1:53–96) in March 1610, written in Latin, the common language of the educated of Europe, and he was able to include his most recent observations from the beginning of that month.

This small book of only fifty-seven pages was dedicated to Cosimo de' Medici, who had become the grand duke of Tuscany with the name Cosimo II after the death of his father, Ferdinando. It was the second book dedicated by Galileo to Cosimo, the first one being *The Function-ing of the Geometrical and Military Compass* (1606). Although he was teach-ing at Padua, Galileo had certainly not forgotten Florence, where he went regularly during the summer vacation period. From at least 1605 on he had been the guest of the Grand Duchess Christina of Lorraine as a teacher of mathematics to her son Cosimo and this gave him the opportunity to improve his relationship with the ducal palace, which had not always gone so well. He especially wanted to deepen his rela-tionship to Cosimo, which the future will show was a wise move.

The first observations reported in the *Starry Messenger* were those of the Moon showing that it had mountains like the Earth. It was not strictly speaking a discovery but a confirmation of opinions held since ancient times, as noted by Kepler in his *Conversations with the Starry Mes-senger* (Galileo, *Opere*, 3.1:112–17). The year before Kepler had tried to describe the lunar geography in his *A Dream, or the Astronomy of the Moon,* a book that today we would consider science fiction but that has many important scientific insights.

Next there was the sensational report of an enormous number of stars only visible with the telescope, including most importantly the Milky Way, which had been a source of wonder since ancient times. Many heavenly clouds were resolved into myriads of stars. But the most important discovery in Galileo's own estimate was that of the four small bodies orbiting Jupiter. Here is what he wrote:

> We have, therefore, a valid and excellent argument for removing any doubt that those might have who, while calmly accepting in the Copernican system the revolution of the planets about the Sun, are very disturbed by the motion of the Moon about the Earth while together they complete a revolution about the Sun each year. For this reason they think that such a structure to the universe should be rejected. (Galileo, *Opere*, 3.1:95)

These words explain why Galileo attributed such importance to his discoveries. Although we know that he had for many years been con-vinced that the Copernican system was the more probable one, he had not yet found the physical basis to justify his conviction. Now finally the telescope opened the way to possible proofs from "sense experience" to show that on two points the traditional cosmology could no longer be upheld. The first was the Aristotelian insistence on an essential difference between heavenly bodies, including the Moon, and Earthly bodies. The existence of mountains and also perhaps, as Galileo suspected, of seas on the Moon showed that the Moon was composed of the same material as the Earth. The second point of dogma was that the Earth was the one and only center of all heavenly motions. The discovery of the satellites of Jupiter showed that there were motions of some heavenly bodies about centers other than the Earth. Of course, this discovery did not decide between the system of Copernicus and that of Tycho Brahe. But now as always Galileo shows no inclination whatever to the system of the Danish astronomer, which he always looked upon as a mere mathematical compromise with no possible physical basis.

The words of the dedication of the *Starry Messenger* to Cosimo II confirm that the discovery of Jupiter's satellites was the principal factor that persuaded Galileo that the system of Copernicus was the true one. While mentioning their high velocity about Jupiter he had, in fact, added that the satellites with Jupiter in twelve years circle the Sun, the "center of the world" (Galileo, *Opere*, 3.1:56).

Inspired by these discoveries Galileo from that moment on had the idea of writing a much more detailed work, his "structure of the world." It would offer a comprehensive view of the reasons favoring the system of Copernicus over all others (Galileo, *Opere*, 3.1:75). In the *Starry Messenger* Galileo promised that this work on the "structure of the world" would appear soon. In fact, it will only be published twenty-two years later as the famous *Dialogue Concerning the Two Chief World Systems*.

The search for proofs of Copernicanism will from now on become Galileo's project for life. He was certainly aware of the vastness of the task before him. Would he be able to carry it out while still fulfilling his teaching duties at Padua? For sure the doubling of his salary and appointment as professor for life provided him with a financial security he never had before. But he would need all the time he could get to carry out his life's program.

There was another motive, which may have escaped most Galilean scholars, for discontinuing his teaching at the University of Padua. It would have become psychologically difficult, if not impossible, for him to continue teaching the old Ptolemaic cosmology when he was now certain that it could no longer be upheld. Should he then teach Copernicanism? But, since he had no new philosophy of nature to replace the old one, such teaching would have put him in open opposition to the Aristotelians like Cremonini. Also, in addition to needing time, he needed the peaceful circumstances required to carry out his research in a thorough way and to freely express his ideas. Polemics with the Aristotelians at Padua would not provide that calm.

In that frame of mind he must have spontaneously thought of returning to Florence. The circumstances were appealing. Upon the death of Ferdinand I just a little more than a year before, his eighteen-year-old son, Cosimo II, an admiring disciple of Galileo's, had succeeded to the dukedom. Galileo had the idea of requesting of him a position as mathematician for life, possibly with the same stipend that he had from the Republic of Venice, but without any teaching obligations. This idea must have come to him at the time of his writing of the *Starry Messenger,* and that is why he dedicated that work to the young grand duke and why he also named the satellites he had discovered about Jupiter the "Medicean Planets." Together with a copy of his book he also sent Cosimo II the telescope with which he had made his discoveries. And he promised to pay a visit to the grand duke in Florence so as to show him "infinitely stupendous things" (Galileo, *Opere,* 10:302). In fact, Galileo paid the visit to Florence during the first days of April, proceeding then to Pisa where the grand duke was at that time. For sure this visit, in addition to the other acts of homage mentioned, contributed to strengthening the bond of friendship between the master and his young disciple. So in Galileo's mind there was no need to hesitate.

When he returned to Padua he sent on May 17 a long letter to the grand duke's secretary of state, Belisario Vinta, in which he explained the reasons why he was inclined to give up his teaching at Padua, the main one being so that he could dedicate his time to research. He gave an account to Vinta of the results he had obtained thus far and of his future projects. About these projects he wrote:

The works that I have to complete are mainly two books on the system or constitution of the universe, an immense idea full of philosophy, astronomy, and geometry; three books on local motion, a completely new science, since no one else neither in times gone by nor in modern ones has discovered the marvelous effects that I demonstrate to exist both in natural and in violent [forced] motions, so that I with reason can call it a new science and founded by me based on first principles; three books on mechanics of which two present demonstrations of principals and fundamentals and one of them on the problems. (Galileo, *Opere*, 10:351–52)

To carry out this program Galileo requested to be absolved from teaching and asked that he be given the title of philosopher in addition to that of mathematician. The dealing between the court of the grand duke and Galileo went ahead rapidly and on July 10, Cosimo II named Galileo "First Mathematician of the University of Pisa and First Mathematician and Philosopher of the Grand Duke of Tuscany." The appointment was for life and carried a salary of 1,000 scudi a year with no obligation to teach or to reside at Pisa. Galileo, sure of his new appointment, had already on June 15 tendered his resignation from the chair at Padua. The Republic of Venice tried to retain him and offered him another substantial increase in salary. But it was of no use. Galileo left Padua on August 2, and he would never return there despite promises made to friends and their repeated invitations to him. He surely was aware that his brusque rupture of relationships with the Venetian Republic had offended the authorities there, since they were convinced that they had always treated Galileo in the best manner.

Copernicanism and the Bible

The start of Galileo's telescopic observations was not the only significant event of the year 1609. In that same year Kepler published his *New
Astronomy,* in which he derived, from the numerous precise observations by Tycho Brahe, an elliptical orbit for Mars with the Sun located at
one of the foci. He later extended this result to the orbits of all of the
other planets. This is known as Kepler's first law. He then formulated a
second law which described the variation in a planet's velocity as it orbited the Sun. In later works, the *Harmony of the World* (1619) and *Epitome of Copernican Astronomy* (1617–21), he formulated a third law, which
relates the orbital period of each planet to its distance from the Sun.

By doing away with the dogma of circular orbits Kepler brought
substantial improvements to the Copernican system. On the basis of
his first two laws, epicycles, deferents, equants, and similar geometric
constructions became superfluous and the simplicity of the Copernican system finally became lucidly clear. On the other hand, without the
development of a new dynamics it was impossible to grasp the full significance of Kepler's laws. The result was that most of Kepler's contemporaries, including Galileo, did not note the importance of Kepler's laws, at least until the publication of the *Rudolphine Tables* (1627). It
would take another forty years before Kepler's laws become widely

known and another twenty before they would be applied to the formulation of a new dynamics in the *Mathematical Principles of Natural Philosophy* of Isaac Newton.

In contrast to the *New Astronomy* of Kepler, the *Starry Messenger* of Galileo immediately aroused great interest. The first printing of five hundred copies was sold out within a week. Galileo sent a copy to the Tuscan ambassador to the court of the emperor of the Holy Roman Empire, at Prague, with a request to have it read by the Imperial Mathematician Kepler. In fact, the ambassador informed Galileo on April 19 that Kepler had read the book and was very pleased with it. But because his telescope was imperfect, Kepler confessed, he had not been able to verify Galileo's observations (Galileo, *Opere,* 10:318–19). On that same day Kepler himself sent a long letter to Galileo in which at the beginning he lamented the fact that he had not heard from Galileo in a long time and that he had received no comments from him on the *New Astronomy.* Nevertheless, he said that he was convinced that Galileo's observations were true and that the conclusions he drew from them were justified. But he was cautious in the way he expressed himself since he had not been able personally to verify the observations (Galileo, *Opere,* 10:319–40). A little later Kepler published that letter with some modifications as *Conversations with Galileo's Starry Messenger.*

Although prudent, the position taken by Kepler was clearly favorable to Galileo. But there were astronomers who were hostile, among them the most well-known Italian astronomer of that period, Giovanni Antonio Magini, who taught mathematics at Bologna. On the way to his return from Florence Galileo had stopped on April 24–25 at Magini's home with the obvious intention of winning the support of this influential professor. Unknown to Galileo was the fact that just before his visit Magini had sent a very negative judgment on the *Starry Messenger* and on the telescope to the Cologne elector (Galileo, *Opere,* 10:345). Magini had also written at almost the same time to Kepler asking his opinion about the satellites of Jupiter (10:341). Kepler answered him on May 10 and sent him a copy of his *Conversations with Galileo's Starry Messenger* with this pointed comment:

> Take it and excuse me. We [Kepler and Galileo] are both Copernican. Like attracts like. But I think that, if you read it carefully, you

will note that I have expressed myself cautiously and have reminded Galileo to stick to his own principles. (10:353)

But Magini could not agree with Kepler's discussions. In his reply to Kepler he stated that the attempt by Galileo, during his stopover in Bologna, to show the satellites of Jupiter "to more than twenty educated persons" at his home was a failure (10:359). Magini continued his campaign against Galileo and warned the most renowned mathematicians of Europe that his discoveries were "pretentious."

The Bohemian Martin Horky, who while studying medicine at Bologna was a guest at Magini's home, was also openly hostile to Galileo. He too sent Kepler the news of Galileo's failure at Bologna and added some nasty remarks about Galileo's physical appearance. A little later he published a small work, *Some Brief Remarks against the Starry Messenger* (Galileo, *Opere,* 3.1:129–45), in which he attacked Galileo and denied the veracity of his observations. The tone of his remarks even disturbed Magini, who threw him out of his home. Even Kepler, upon receiving from Horky a copy of his little work, severely condemned his way of acting and suggested that he go back to his homeland (10:419). But Horky had his work distributed in Italy and in other European countries. A copy was read in Florence by Ludovico delle Colombe, who, as we shall see, will become one of the leaders of the faction against Galileo in that city. Another wound up in the hands of Francesco Sizzi (1585?–1618), who at that time was writing a book against Galileo's discoveries.

Negative reactions from the Aristotelians also began to surface quickly. At Padua the best known professors, including Cremonini, refused outright to look through the telescope even though Galileo had invited them "an infinite number of times" to do so, as he wrote on July 19 in a response to Kepler (10:423). But he added that many others had verified the reality of his observations. On August 30 Kepler was finally able to use the telescope that Galileo had sent the Cologne elector. After many accurate observations he became fully convinced of the truth of Galileo's discoveries and he made this public: "A recording of my own observations of the four satellites of Jupiter which the Florentine mathematician Galileo Galilei by right of having discovered them has named the Medicean Stars" (Galileo, *Opere,* 3.1:181–88).

This support by an astronomer of Kepler's class and the spread of the telescopes made by Galileo began, as time went on, to attenuate reservations and opposition among scientists. Thus his friend Martin Hasdale could write to him towards the end of 1610 from Prague that his discoveries were no longer contested there and that even Horky was by now convinced that he had been in error and bitterly regretted that he had published the book against Galileo, thereby compromising his own reputation.

Just before his departure from Padua, Galileo made another discovery, this time concerning Saturn. He referred to it confidentially in a letter of July 30 to Secretary Vinta:

> the star of Saturn is not alone but is composed of three which almost touch one another, nor do they ever move or change and they are placed in a line along the length of the zodiac, the one in the middle being three times greater than the other two. (Galileo, *Opere,* 10:410)

Galileo made an even more important discovery when, about two months after his arrival in Florence, he observed the phases of Venus. This could in no way be made to agree with the astronomy of Aristotle or of Ptolemy. It was, in fact, a proof that Venus went around the Sun; it was a phenomenon consistent only with both the system of Copernicus or that of Brahe. But, as we know, for Galileo the system of Brahe had no physical basis, and so for him the discovery of Venus's phases was an incontrovertible first proof of heliocentrism. This conviction is evident from a letter sent by him to the Tuscan ambassador at Prague on December 11, 1610 (Galileo, *Opere,* 10:483).

At the end of 1610 Galileo received another valuable support for his discoveries from the Jesuits of the Roman College and, in particular, from Clavius. Previously they had been skeptical, and even Clavius had not been convinced, though Galileo had sent him a letter in September giving him the details of his observations (10:431–32). But towards the end of November Clavius began to reconsider his doubts (10:479–80); soon afterward his hesitancy completely vanished. On December 17 he so informed Galileo, and, after having told him that he himself and the other Jesuits of the Roman College had seen the Medicean Planets very clearly, he added: "Truly Your Lordship deserves high praise having been the first to observe this" (10:484). Clavius also informed Galileo that at the Roman College they had not been able to observe the two little stars attached to Saturn that Galileo had described but that they had only seen Saturn to be "oblong." Clavius encouraged Galileo to continue his observations which might lead him "to discover new things about the other planets." As to the Moon, he confessed: "I marvel very much at its unevenness and roughness, when it is not full" (10:485).

Galileo was obviously very pleased also with this letter and he replied thus on December 30:

> Your Reverence's letter has been all the more gratifying to me since, although I much desired it, I did not expect it from you. When it arrived I was rather indisposed and almost confined to bed and it helped in no small part to pick me up from my illness since your letter won for me a testimony to the truth of my new observations and gained to my cause some of the incredulous; but there are still those who persist in their obstinacy and regard your letter to be either fictitious or written only to please me, and they are, to put it briefly, waiting for me to have at least one of the four Medicean Planets brought down from heaven to Earth so as to give an account of itself and clear up any doubts. (10:499)

Galileo added the news of his discovery of the phases of Venus and insisted on the extremely important consequences of it:

> Now look here, my Sir, how clear it is that Venus (and doubtlessly Mercury will do the same) goes about the Sun, the center without a doubt of the major revolutions [orbits] of all of the planets. Furthermore, we are certain that those planets are of themselves dark and they are only bright by being illuminated by the Sun (this effect does not occur, I think, for the fixed stars from what I have been able to observe); and this system of planets is surely different than has been commonly held. (10:500)

From remarks he made in this letter, Galileo's health was not good at that time. In January he was able to go to the villa in the countryside of his former disciple from Padua, Filippo Salviati (1582–1614), and thus to get some relief from the humid winter in Florence. While there he received a letter from the Dominican Tommaso Campanella (1568–1639), who at that time was in prison in Naples for his political activity against Spain. Campanella had read the *Starry Messenger* and praised Galileo for having "purged mankind's eyes by showing them a new heaven and a new Earth on the Moon" (11:23). He thus made it clear that he had intuited the revolutionary nature of Galileo's discoveries and the use to which they could be put, at least against the world of Aristotle and Ptolemy. But, despite his admiration for Galileo, it does not seem that Campanella ever embraced Copernicanism. Galileo did not respond to him and would in the future take a rather reserved position with respect to the Dominican thinker with whom he did not agree for sure on his cosmological ideas.

While in this way the opposition to science was slowly abating, the same cannot be said for philosophical issues. And, tightly linked to philosophical considerations, theological concerns began to appear. One of Galileo's friends in Padua, the priest Paolo Gualdo, informed him in May 1611 that Cremonini was intending to write a refutation of Galileo's claims but without naming him (11:100). Cremonini's gesture of leaving Galileo's name out of the work was clearly out of respect for his friendship with Galileo despite their divergent ideas. Gualdo added some advice for Galileo:

> As to the matter of the Earth turning around, I have found hitherto no philosopher or astronomer who is willing to subscribe to the opinion of Your Honor, and much less would a theologian wish to do so; be pleased therefore to consider carefully before you publish this opinion assertively, for many things can be uttered by way of disputation that it is not wise to assert, especially when the contrary opinion is held by everyone, imbibed, so to speak, since the foundation of the world. . . . It seems to me that you have acquired glory with the observation of the Moon, of the four Planets, and similar things, without taking on the defense of something so con-

trary to the intelligence and the capacity of men, since there are very few who know what the signs and the aspects of the heavens mean. (11:100–101)

As a priest Gualdo was particularly concerned about the reactions from theologians to a position by Galileo that was too explicitly in favor of Copernicanism. This concern was clearly justified in light of the events that followed. In fact, as the possibility of refuting Galileo's discoveries on astronomical grounds or on those of natural philosophy slowly began to weaken, the opposition to them on theological grounds would begin to increase and would play a decisive role as the drama of the "Galileo Affair" unfolds towards a conclusion. We must, therefore, consider the reasons behind this opposition.

The tension described was certainly nothing new. The problem of how to relate the biblical depiction of the world's structure and the motion of the heavenly bodies to the affirmations of natural philosophy and astronomy existed, in fact, right from the beginning of the spread of Christianity into those regions of the Roman Empire that were most directly in contact with Greek-Hellenistic culture. Having emerged from the Jewish religion, the new Christian religion had kept the patrimony of the faith contained in the Bible, the sacred books of the Jewish people. At the beginning of the Bible, in the book of Genesis, Christians found an account of the creation of the world and a description of it that were repeated in many other texts of the Bible. Overall it was a primitive view dependent very directly upon the cosmology and partly on the cosmogony of the Babylonians. A flat Earth was surrounded by oceans, and beyond, at the edges of the Earth, lofty mountains arose and held up the solid vault of the heavens or the "firmament." Under the Earth the "lower waters" spread out to form the rivers and seas and still deeper down was the realm of the dead. Above the vault of the heavens were the "higher waters," which could not pour down on the Earth due to enormous "closures" in the firmament. Only in exceptional cases, such as in Noah's time, could the closures be opened, causing a universal flood. The heavenly bodies moved across from east to west in the space between the Earth and the firmament without any explanation being given of their daily appearance and disappearance. So it was

a depiction of the world in sharp contrast to that given by the philosophy and mathematics of the Greek-Hellenistic culture. For educated
pagans who had become or wished to become Christian, the problem
of the relation between these two conceptions of the world could not
be avoided.

One possible solution was that of the intransigents, Tertullian
(ca. 200 CE) and a bit later Lactantius (240–320 CE). They rejected
any possibility of compromise between the Christian faith, founded on
the literal sense of scripture, and the Greek-Hellenist culture. Thus they
denied the conclusions reached by philosophers, mathematicians, and
astronomers. A more positive and conciliatory position sought to distinguish between the true faith as expressed in the Bible in ways that
were not scientific, and so could not be in contrast to the conclusions
from natural philosophy and astronomy of that epoch. This attitude
first came to light among the thinkers of the school of Alexandria and
was then taken up and deepened, with considerable differences, by the
Greek and Latin Church Fathers. Special mention should be given to
the greatest of the Latin Fathers, St. Augustine (354–430 CE). His interest in philosophy in general went through a progressive transforma-
tion after his conversion (387 CE) and became extremely critical. But
his view of natural philosophy and astronomy remained relatively positive, except from some reservations about the risk that astronomy be
used for astrological purposes. But what was one to think when the conclusions from Greek scientific thinking appeared to contradict Holy
Scripture? In his writings, especially in his *De Genesi ad Litteram,* he provides a general answer to this question. He gives importance to the fact
that "one does not read in the Gospels that the Lord said: I will send
the Paraclete so that he may teach you the course of the Sun and the
Moon. Because he wanted to make them Christians, not mathematicians." And he adds that the sacred writers had no intention to teach
anything about the form and figure of the heavens nor about any questions about nature "since such knowledge was of no use to salvation"
(*De Genesi ad Litteram* 2.9).

This call for prudence is repeated many times in Augustine's writings. He warns of the risk of making hasty judgments on statements
from Greek "science" based on texts from the Bible, since one day such

statements might be demonstrated to be true. And so derision could be cast on the scriptures by nonbelievers and the way to salvation would be closed to them. But in Augustine's writings there is also a dominant theme of preferring "those matters that concern salvation" to the search for an understanding of nature. There is no doubt that Augustine shows no hostility in principle to Greek "science" and, in fact, is prepared to recognize its sphere of validity. He simply thinks that possessing such knowledge is not necessary for a believer. A Christian could dedicate his intellectual energies with much greater spiritual good to the matters of the faith.

All of these tendencies maturing during the period of the Fathers could not for the moment be deepened because of the profound cultural crisis created by the continuing breakup of the Western Roman Empire and the almost complete loss of contact with the Eastern Empire, that of Byzantium. A major negative impact was the loss of the knowledge of Greek combined with the great scarcity of Latin translations of the treasures of philosophy and of Greek-Hellenistic science. And so the knowledge of these treasures almost completely disappeared in Western Europe during the early Middle Ages.

The rediscovery of the cultural patrimony of the Greeks began in the twelfth century. A fundamental role in the process was played by the philosophical assimilation of Greek thought in the Arab world during the previous centuries. The Greek texts in philosophy translated into Arabic and the Arabic commentaries on them became known to the European scholars through their contacts with Arabic cultural centers in Spain and in Sicily. This was followed by an intense work of translation of these Arabic texts into Latin and later on directly from the Greek, signaling the beginning of an intense work in the study and interpretation of Greek and Hellenistic thought. This work took place especially in the "schools," the medieval universities, from which we have the name "scholastic," given to the great intellectual movement born in those schools.

The position taken towards Greek philosophy by the principal exponent of scholasticism, St. Thomas Aquinas (1224–1274), is totally different than that of St. Augustine. For Aquinas, to be involved in philosophy was certainly not a "waste of time." As a theologian he saw in

Greek philosophy a necessary rational preface to the Christian faith. Certainly it was necessary to "cleanse" those philosophical conclusions that were in contrast with the Christian faith, such as the eternity of the world or the mortality of the soul. But once so purified Greek philosophy was useful to establish the rationality of the premises, such as God's existence, on which the theological affirmations of Christianity were founded. But, in addition to the usefulness of philosophy as "the servant of theology," Aquinas had an interest in philosophy in general, including natural philosophy, as shown by his great commentaries on Aristotle's *Physics* and *On the Heavens,* where he shows his allegiance to Aristotle's physics and cosmology. But that allegiance forced Aquinas, as it did other medieval scholastics, to face the problem of interpreting the description of the world given in Genesis and elsewhere in scripture. Following Augustine, Aquinas repeats the fact that scripture frequently uses expressions to fit the common understanding of people of that age and so they need not be taken literally. Even though he would accept that every word of the Bible is true, as inspired by God, there are cases in which such truth need not be derived from the common and immediate meaning of the words.

A very important comment is required here. Similarly to the other scholastics, Aquinas does not use faith as a higher criterion to force agreement with the conclusions of Greek natural philosophy. On the contrary, he is open-minded enough to accept those conclusions once they are certain and to reexamine the traditional interpretation of the creation accounts and other biblical passages in light of those conclusions. In particular, his preference for the geocentric theories of antiquity did not arise at all from his wish to reconcile scripture with natural philosophy. In fact, the only thing known about the Earth's motion in Aquinas's time were the theories of Philolaus and Heracles about its axial rotation, and the only time that Aquinas mentions Aristarchus is where he takes his theory as the same as that of Heracles. So the only reason that Aquinas rejects an axial rotation of the Earth is because it is "philosophically absurd" according to Aristotle. There is no doubt that Aquinas would have accepted that the Earth moved had such a theory, as well as Aristarchus's heliocentrism, been proposed as true by the Greeks; and Aquinas would not have seen it to contradict scripture.

This is confirmed by the fact that he accepted Aristotle's cosmology, which agreed with scripture on only one point, the immobility of the Earth. On all else the primitive biblical view, including the very idea of a flat Earth, was separated from the Aristotelian view by an abyss. If Christian thinkers had the courage to cross over this abyss it was because they accepted without question the cosmology of Aristotle. This secure position in natural philosophy provided them the firm basis to proceed without taking an intransigent position on the literal sense of scripture. To conclude, Aquinas and the other medieval thinkers accepted the Greek-Hellenistic geocentrism as true because it was seen as such by the greater majority of "pagan" thinkers.

The notion of an axial rotation of the Earth, however, was considered by various scholastics, at least as a theoretical possibility. Of particular interest are the statements of Jean Buridan (1295?–1358) in his *Questions on the Heavens and on the World* and the even more developed considerations of Nicole Oresme (1325–1382) in his *Book on the Heavens and on the World*. Oresme examines especially the traditional arguments against any motion of the Earth, including those from experience, from natural philosophy, and from scripture, and he shows that they are not conclusive using reasons that anticipate in some way those that will be offered by Copernicus and Galileo. Oresme then goes on to examine the arguments in favor of the Earth's rotation and, in agreement with Buridan, he first invokes the principle of relative motions. And so he proposes that the daily motions of the heavenly bodies could be explained either by the fact that they are actually moving or by the Earth's rotation. His thought, like that of Buridan, is that it is simpler to have the Earth rotate than to have the sphere of the fixed stars rotating at an enormous speed and dragging along with it in twenty-four hours all of the other spheres. It is more reasonable, he continues, to have the Earth, which is made up of the four less noble elements, move and to have the heavens, made of the noblest element, the ether, remain stationary. Rest is more noble than motion. But, despite all of these arguments, Oresme concludes:

> And still they maintain and I [also] believe that the heavens move and that the Earth does not: because "God fixed the Earth in a

way that it could not move" (Psalm 92); [and this] despite reasons
to the contrary, because they constitute *persuasive* arguments *which
do not evidently prove. (Livre du ciel et du monde* 2.25; emphasis added)

Even though he was more detached than Aquinas from the natural
philosophy of Aristotle, Oresme was aware that arguments supporting
the Earth's rotation were insufficient, and, in the absence of convincing proofs, he preferred to draw upon the literal meaning of the Bible.
This position was assumed by many of Galileo's contemporaries who,
less intransigent than the dyed-in-the-wool Aristotelians, would not exclude a priori the possibility of clear proofs of the Earth's motion. But
in the absence of, perhaps even hoping for, proofs, they would choose
a prudent and "pious" allegiance to scripture.

During the three centuries between the philosophical-theological
synthesis of Aquinas in the *Summa Theologiae* and the *On the Revolutions*
of Copernicus, scholasticism began little by little to lose its creativity
and to become dedicated simply to commentaries on its medieval authors. On the other hand, despite the occasional attack against it, Aristotle's philosophy maintained the privileged position that the Thomistic
synthesis had won for it, even to the point where it seemed to be an essential component of the Christian view of the world. Such a view was
particularly dominant among Dominicans, the religious order to which
Thomas Aquinas belonged and that had traditionally accepted the role
of defending Catholic orthodoxy. Such a conviction of an obligation
to defend a philosophical-theological patrimony whose validity was in-
disputable could not but create a climate of strict closure in complete
contrast to the openness of the broader medieval culture. In fact, as we
will see, among Catholics various Dominicans will be among the first
critics of Copernicus as well as the most determined in their opposition to Galileo.

In addition to the deprivations suffered in scholasticism another
profoundly negative occurrence, which had an effect upon Copernicus
but even more so on Galileo, was the emphasis placed on the literal
meaning of the Bible as a result of the bitter polemics in theology between Catholic and Protestant at the time of the Reformation. It was
as a matter of fact by their insistence on the literal meaning of scripture

that the reformers mounted their attack on the Catholic Church. According to them the hierarchical structure of the Church and especially the papacy and the authority attributed to it could not be justified from a literal reading of the New Testament. For their part the Catholics sought to show on the contrary that it was precisely an authentic literal reading of the Bible that established the characteristics of the Roman Church and in particular the papal authority. This heated controversy over the literal reading of the Bible could not but have serious repercussions upon Copernicanism. In fact, even before the publication of *On the Revolutions,* the first negative reactions to Copernicanism came from Protestants armed with their literal interpretation of scripture. Martin Luther himself (1483–1546), the principal protagonist of the Reformation, in the course of a convivial conversation had declared: "That mad man [Copernicus] wants to turn astronomy on its head, but, as scripture says, Joshua commanded the Sun and not the Earth to stand still" (Luther, *Tischreden,* 2:419). A close collaborator of Luther, Melanchthon (Philip Scharzerd called Melanchthon, 1497–1560), who was a dominant figure at the University of Wittenberg, likewise opposed Copernicus and continued to do so until a number of years after Copernicus's death. Undoubtedly, the reason why Osiander, a Protestant, had inserted his "Notice to the Reader" in *On the Revolutions* was principally to remove the possibility of future negative reactions from Protestants. But even though this expedient had some influence among Protestant theologians, it did not dissuade those who were better informed, including Melanchthon. Nonetheless, the criticisms of Copernicus by the reformers never grew to become an official position of the Protestant Church. That was certainly due to the fact that within Protestantism there was no Church structure with a centralized authority such as in Catholicism. But what was Catholicism's initial reaction to Copernicanism?

In the past Catholic writers on Galileo generally held that the attitude to Copernicanism among Catholics was for many decades directed towards moderation, even partiality. In support of this thesis there is before all else the invitation extended to Copernicus to contribute to the reform of the Julian calendar. The reform project was announced by Leo X at the Fifth Lateran Council (1512–1517) and Copernicus was among the astronomers "of renowned fame" invited to participate.

This invitation is recalled by Copernicus in his dedication of *On the Revolutions* to Paul III. In fact, between 1511 and 1513 Copernicus had distributed the manuscript of his *Commentariolus,* a first sketch of his system. This contributed to a wider knowledge of his ideas and probably also to his "renowned fame." Thus it does not appear that there were at that time any suspicions of his ideas being contrary to scripture.

In 1533 Pope Clement VII received in the Vatican gardens "an explanation of the opinion of Copernicus on the movement of the Earth" presented by Johann Albrecht von Widmanstadt (1500–1577), whose information on Copernicus's theory probably came from Theodoric of Radzyn, the representative to Rome of the Chapter of Warms, to which Copernicus also belonged as a canon. This garden explanation is noted by Widmanstadt himself in a precious Greek manuscript given to him in gratitude by the pope, an indication that the latter was pleased with the explanation. After the death of Clement VII, Widmanstadt went into the service of Nicholas Schönberg (1472–1537), a confidant of the deceased pope, who was appointed cardinal by the succeeding Pope Paul III (1534–1549). Influenced by Widmanstadt, Schönberg took an interest in Copernicus's ideas and asked Theodoric of Radzyn to have copies made at Frombork, where Copernicus resided, of all of Copernicus's writings and to have them sent to him at Rome. In a letter to him in 1536 Schönberg himself exhorted Copernicus to publish his writings. Copernicus inserted this letter at the beginning of the *On the Revolutions.* A like exhortation to publish came to Copernicus from his old colleague, the canon Tiedemann Giese, who afterward became bishop of Chelmno. After the *On the Revolutions* was printed, Giese raised a severe protest that the "Notice to the Reader" had been inserted, because it was a betrayal, in his view, of the real intentions of the author, which he had always shared.

Finally there is the fact that Copernicus himself dedicated his work to Paul III, thus trying to protect it from being attacked by "calumniators," as he wrote in the dedication. This seems to indicate that Copernicus trusted in the open-mindedness of the pope. And, in fact, the Catholic Church took no actions against the *On the Revolutions* then nor in the period after that leading up to 1616. Recent studies, however, indicate that this silence of the Church should not be interpreted as a sign of an impartial suspension of judgment on Copernicanism. In fact, testimony to the contrary is given by the Dominican Giovanni Maria Tolosani (ca. 1470–1549) in an appendix to his treatise *On the Most Pure Truth of Holy Scripture,* which, though finished in 1545, remained in manuscript. Tolosani belonged to the monastery of St. Mark in Florence. So did Tommaso Caccini, another Dominican who many years later had the opportunity to read that manuscript as testified by a marginal note he left there. He would become one of the fiercest adversaries of Galileo.

According to Tolosani, the Master of the Sacred Palace Bartolomeo Spina, also a Dominican, had decided to condemn the work of Copernicus but died before he could do so. We can probably trust this information because Tolosani and Spina were close friends. The successor to Spina had neither the time nor the desire to take up the issue. Nor could Pope Paul III get involved. Although he may have received Spina's negative view of Copernicus, he was surely completely absorbed in the grave problems of that time, which were to be addressed by the Council of Trent, which was just beginning its work. Even if he were aware of the Copernican issue, it could not compare to the other important and urgent problems that he faced. For the same reasons the Church remained silent about Copernicanism for the next seven decades. On the other hand astronomers had not taken up the cause in any significant way. So for the Church there was no real danger to be faced with the task of making doctrinal pronouncements.

The attitude of Catholic theologians of that time confirms that the silence of the Church cannot be interpreted as an acceptance of Copernicanism. There were, of course, some who lined up with Copernicus, among them Giese, along with the Protestant Rheticus, who convinced Copernicus to publish. The two of them were well aware of the objections from scripture, and they wrote in response to those objections. Giese's response has been lost, but that of Rheticus, published in a seventeenth-century book, has recently been found. Further support for Copernicanism is found in Diego de Zuñiga's commentary on the Book of Job, published first in Toledo, Spain, in 1584 and then in 1591 in Rome. The Augustinian de Zuñiga, professor of theology at the University of Osuna in Spain, examined in his commentary the biblical text of Job 9:6, "He [God] shook the Earth and moved it from its place causing its pillars to tremble," and concluded that it was much easier to interpret the passage by following the opinion of the Pythagoreans that "in our age Copernicus has demonstrated." Those passages of the Bible that speak of the Sun moving simply reflect, according to de Zuñiga, the common way of speaking, used even by Copernicus and his followers; in fact, it is the Earth that moves. As we shall see, this book will be involved in the actions taken by the Catholic Church in 1616 against Copernicus's theory, but before that hardening of the Church's position, it does not appear that de Zuñiga's opinion created any problems for him, although it was criticized by the Jesuit theologian Juan de Pineda in his commentary on the Book of Job (1597–1601). Perhaps as a result of such or similar criticisms, in a subsequent edition of his commentary de Zuñiga explicitly rejects his previous pro-Copernican interpretation.

In contrast to these supporters of Copernicanism, there were many others, including among Catholics, who were hostile. In addition to Pineda there were surely Spina and even more so his friend Tolosani, who in his appendix mentioned above severely criticizes Copernicus. As to philosophy, he accuses Copernicus of the most crass ignorance of the arguments from natural philosophy already proposed by Aristotle against the Pythagorean theory of the motion of the Earth; and as to theology, he states that the Copernican theory "is opposed to Holy Scripture which declares that the Heavens are on high and the Earth down low." Tolosani concludes with a warning that unfortunately will prove to be prophetic: "The Pythagorean theory [Copernicanism] could easily give rise to quarrels among the Catholic interpreters of Sacred Scripture and those who would obstinately adhere to a false belief. I have written this little work in order to avoid such a scandal."

A negative judgment, although more moderate, was given quite a bit later by the Jesuit mathematician Clavius, who played an important part in the reform of the calendar that was finally promulgated in 1582 by Gregory XIII (1572–1585). In his *Mathematical Works* (1611–1612) Clavius shows a great respect for Copernicus as a mathematician. But he adds that "many errors and absurdities are contained in his position." After giving some examples he concludes: "All of this is contrary to the common teaching of the philosophers and of the astronomers and it appears to contradict what is taught in many places in the Holy Scriptures."

Although there was no official position, the conviction within the Catholic Church that Copernicanism could not be reconciled with scripture comes to light clearly in the tragic affair of Giordano Bruno (1548–1600), who at the age of seventeen entered the Dominicans. He left the order in 1576 to avoid being tried for his ideas, which were considered heretical by his superiors. He wandered around for many years through Switzerland, France, England, and finally Germany; he was almost always at the center of controversies for his philosophical and theological ideas. Finally in 1591 he returned to Italy and settled in Venice at the invitation of the patrician Giovanni Mocenigo, possibly hoping to obtain the chair of mathematics at Padua, which was given to Galileo a year later. In 1592 he was accused by Mocenigo himself of holding heretical ideas and was subjected to a first trial by the Inquisition in Venice. A year later he was put into the hands of the Roman Inquisition. After a new trial that lasted seven years Bruno in the end refused to retract and was condemned to be burned at the stake as an impenitent and pertinacious heretic. The acts of Bruno's trial have been lost, due probably to all that happened with the transfer of the Roman archives to France under Napoleon. Still from all that can be pieced together of the documents that remain, especially the "Summary of the Trial," composed in March 1598 (a copy for the use of the assessor of the Holy Office, Marcello Filonardi, is extant), we are able to sufficiently reconstruct the reasons for this condemnation.

Bruno was inspired by neoplatonic philosophy and by the ideas of Cardinal Nicholas of Cusa (1401–1464) of an unbounded universe. He affirmed that there was an infinite universe endowed with an infinite dynamism, as the "visible image" of the infinite God. He never took up a clear position of "monism" by identifying this natural world with God, and throughout his trial he denied that he had ever done so. But his judges concluded that, in fact, he was a monist and so a dangerous heretic. This is without a doubt the fundamental reason, a philosophical one, for Bruno's condemnation. But there were also theological considerations. He was accused by Mocenigo and other Venetians of denying the dogma of the Trinity by identifying the Holy Spirit with "the soul of the world," the source of its dynamism. He was also accused of denying the divinity of Christ and almost all the other Christian dogmas. But Bruno had always denied that there was any foundation to such accusations, so that in the end they could not be proved. There will remain only Bruno's "obstinate impenitence" as to his philosophical ideas, which, at the end of his trial, will provide proof also of his errors in theology. And so these mistaken ideas in theology will be inserted as a whole in the sentence of condemnation, of which only a partial copy lacking the list of accusations (except for that of transubstantiation in the Eucharist) has been saved. But nine days after the execution of Bruno, fourteen of the articles of condemnation were announced from memory by the converted Protestant Kaspar Schoppe who had witnessed in person the reading of the sentence of condemnation. Even though most of those articles concern Bruno's theological heresies, the fifth one is about his philosophical idea of the existence of innumerable and eternal worlds.

Bruno had made it clear that his view of material reality was based upon Copernicus's theory, but he had gone well beyond the great Polish thinker, who had mentioned the problem of the infinity of the universe in his *On the Revolutions* but had prudently remarked: "Let us leave it to the philosophers to dispute whether the world is finite or infinite." Thus Copernicus had kept in his book to the idea of a finite world, whereas Bruno, on the basis of the atomism of Democritus and Lucretius, had boldly gone further and claimed that the world was infinite and eternal and that in the innumerable planetary systems therein there were living, intelligent creatures. According to the summary of the trial, during the interrogations he had affirmed to leave the question whether they were like humans to the free choice of those who wanted to think so. Such an (apparently) uncommitted position shows that he was well aware of the significant theological problem that would arise from personally supporting the existence of human-like creatures in the universe. This supposition, in fact, appeared to his judges to be directly in conflict with the Christian doctrine of Christ's redemptive death, which happened "once and for all" (*semel et pro omnibus*). In an infinite universe like Bruno's, would Christ have had to die an infinite number of times? Again from the summary we see that Bruno, pressed by his interrogators to express his *own* opinion on the matter, had tried to escape the danger, affirming, in a somehow confused manner, that such intelligent beings were immortal, and therefore either angels or like Adam and Eve before the original sin, and thus—in both cases—not in need of any redemption. This explanation, however, does not seem to have convinced his judges, as appears (at least indirectly) from the already mentioned inclusion of Bruno's thesis of the infinite worlds among the heresies that had led to his condemnation. Now, as he himself had stressed at the very end of the trial, that thesis, as a purely philosophical one, had never been condemned by the Church. It was indeed because of the contention that they were inhabited that the theory of infinite worlds could have been considered heretical.

But was the Copernican theory of the Earth's motion also involved? It appears in the list of "censures" to which Bruno had to answer during the final questioning period of the trial. We do not know whether this specific censure was in the end included in the eight propositions considered heretical and to which Bruno was required to abjure on January 18, 1598. These propositions had been garnered from the acts of the trial and from the writings of Bruno at the suggestion of one of the consultors of the Holy Office, the Jesuit Robert Bellarmine (1542–1621), whom we shall see again in the first stages of the Galileo Affair. But the very existence of the "censure," whether it was later on included among the eight propositions or not, is a clear sign of the concern of Church authorities for Copernicanism, which they most certainly at that time judged to be contrary to scripture. Bruno had correctly maintained that the "Notice to the Reader" in *On the Revolutions* was not written by Copernicus and that the latter had presented the theory as the way in which the world was really made, a position which only aggravated the concerns of his judges. From Bruno's response to his judges we can gather that he was confronted with the verses *Terra autem stat in aeternum* ("But the Earth stands in eternity") and *Sol oritur et occidit* ("The Sun rises and sets"). These same verses would be used against Galileo, and Bruno's response is along the same lines as that of Galileo. Since there existed no official dogmatic statement of the Church on the matter, there could be in strict theological terms no question of heresy, a point which Bruno made insistently in his defense. But, according to the theologians of the Holy Office, it was clearly contrary to Holy Scripture and the teachings of the Church Fathers and on this basis Bruno was probably asked to abjure, as would happen to Galileo thirty-four years later.

Of course, Bruno, like Copernicus himself, had no physical arguments in support of Copernicanism; in fact, Bruno had only a summary knowledge of the theory. But now, thanks to Galileo's discoveries, the view of Copernicus was no longer a mere astronomical theory or philosophical speculation. It had become a real physical possibility, and, although not proven, certain facts undeniably favored it. There remained, of course, the theory of Tycho Brahe, which like Copernicanism could explain Galileo's discoveries. And now theologians found themselves facing a real problem. How could they admit even the possibility of a theory that contradicted the literal meaning of passages in scripture? It is important to realize that at the beginning of the polemics between Galileo and his adversaries it will be the Aristotelian philosophers who will bring in the arguments from scripture to support their arguments from natural philosophy. The concerns of the theologians were already present, but they would not be given primary consideration until later on in the controversy. Still, the slow but sure movement of the center of gravity from the astronomical and philosophical considerations to those from scripture marks the real beginning of the Galileo Affair.

A first example of this turn of affairs is found in the anti-Copernican actions taken by the Aristotelian philosopher Ludovico delle Colombe (1565–?) at Florence. They were to have serious consequences. Between the end of 1610 and the beginning of 1611, he spread about in manuscript a dissertation of his entitled *Against the Motion of the Earth*, in which, having given the arguments from natural philosophy, he took up objections from scripture quoting at the end the words of the Jesuit theologian Pineda in his *Commentary on the Book of Job IX, 6*: "Others call this opinion, taken up from the mouth of the ancient philosophers from Copernicus to Calcagnino, crazy, senseless, temerarious and dangerous for the faith and more fit to show that one is clever rather than to promote in a useful way the advance of philosophy and astrology." And he added: "Perhaps these pitiful persons [Copernicans] will have recourse to interpretations of the Scripture which do not coincide with the literal meaning? Not at all: because all theologians without exception say that when Scripture can be taken in a literal sense, it should not be interpreted differently." This would become the battle cry as the story against Copernicanism unfolds.

Galileo read the dissertation of delle Colombe attentively and had annotated it. As a manuscript it would have had limited distribution. It contained numerous errors, and there were passing attempts to explain Galileo's discoveries by having recourse to Aristotelianism. But what must have disturbed Galileo most was the dangerous precedent of drawing scripture into the discussion.

Another attack, directed specifically by name against Galileo, came from Francesco Sizzi (1585?–1618), another Florentine who had contacts with Magini and Horky. In his *Dianoia astronomica, optica, fisica,* published in 1611 in Venice, he, like delle Colombe, added arguments from scripture to those from astronomy and physics and declared, among other things, that there could only be seven planets because there were only seven arms on the candelabra in the Jerusalem Temple; therefore, the four Medicean Planets could not exist. In addition to these writings Galileo must have heard on more than one occasion arguments from scripture in opposition to his discoveries, and this only added to his concerns about possible negative reactions from Roman authorities. In his favor was the valuable support of the Jesuit mathematicians of the Roman College, who may not have followed him all the way to Copernicanism but who were perfectly aware that his discoveries were not compatible with Aristotelianism and did not exclude a priori that someday Copernicanism might be demonstrated to be true. And so they were cautious about making hasty pronouncements that Copernicanism was incompatible with scripture, knowing also that they were following in their prudence a long theological tradition from the time of St. Augustine.

These Jesuits were anxious to have Galileo come to Rome. Clavius was nearing the end of his life but still maintained a vivid interest in astronomy. And there were his disciples, especially Christoph Grienberger (1561–1636) from the Tyrol and the Flemish Odo Van Maelcote

(1572–1615), both of whom had taken up telescopic observations and undoubtedly would have liked to compare their data and conclusions with those of Galileo. In fact, Galileo came to Rome at the end of March 1611 and stayed at the Tuscan embassy, located in Palazzo Firenze. A sign of how important he saw his relationship to those Jesuits is the fact that the first visit he made on the day after his arrival was to the Jesuit mathematicians. Right after that he dedicated himself to meeting persons of importance in Church circles and in Roman cultural groups with the intent of convincing them through telescopic observations of the truth of his discoveries.

Without a doubt the impression made by his telescopic observations must have been strong, even if opinions about them remained divided. A proof of this is seen in the fact that the Jesuit Cardinal Robert Bellarmine, one of the most authoritative theologians of his day, felt it necessary to have the opinion of the Jesuit mathematicians of the Roman College on the telescopic observations. On April 19 he wrote them:

> I am aware that Your Reverences have knowledge of the new astronomical observations of a worthy mathematician by use of an instrument called a canon or more exactly a looking glass; and I too have seen several marvelous things about the Moon and Venus. But I wish that you would do me the favor of telling me sincerely your view. . . . I wish to know because I hear contrary views; and Your Reverences, trained in the natural sciences, will be easily able to tell me whether these new results are well founded, or are rather only due to appearances and are not true. (Galileo, *Opere,* 11:87–88)

Bellarmine continued by posing the following precise questions: (1) Were there really a multitude of stars invisible to the naked eye? (2) Was Saturn really composed of three stars together? (3) Did Venus really have phases like the Moon? (4) Was the lunar surface really rough and uneven? (5) Did Jupiter really have four satellites revolving around it?

The reply of April 24, signed by Clavius, Grienberger, Maelcote, and Lembo, confirmed the reality of Galileo's discoveries. Only as regards the fourth question was there a report of a difference of opinion among the consultants. Clavius thought it more probable that the surface of the Moon is not unequal, but rather that the perfectly spherical lunar body is not uniformly dense and that it has parts that are more dense and parts more rarefied. Others thought that the surface is truly unequal; but they were not certain. As we see, we are dealing with a profoundly honest answer, even though one notes in Clavius's case that he was influenced by philosophical worries of an Aristotelian mold that prevented him from accepting unconditionally the discoveries with respect to the Moon, thought to be until then a perfect body.

What were Bellarmine's motives in writing this letter? He met Galileo, as he himself tells us, when the latter visited Rome. Whether the letter was written after their meeting or whether the meeting was imminent, it is totally understandable that Bellarmine would be interested in having the opinion of these Jesuit "experts." But it is probable that Bellarmine had more than a passing interest and that he was seriously concerned. As we have seen, he had already taken part in the trial of Giordano Bruno and, having been made a cardinal towards the end of the trial, had, as a member of the Holy Office, voted for his condemnation. Now the echo of the heated discussions of Galileo with his Roman counterparts had surely reached the Holy Office, and Bellarmine was still one of its authoritative members. The minutes of the Holy Office's meeting of May 17 show this, for at the bottom the following note is added: "Check whether in the trial of Cesare Cremonini mention is made of Galileo, a professor of philosophy and mathematics." As we know, Cremonini was a professor of natural philosophy at the University of Padua and for most of his life he had worked under the accusations of heresy, of denying, for example, the immortality of the soul, and even of atheism. These accusations came to the attention of the Holy Office and from 1604 on there were a series of trials and replies of self-defense by Cremonini, resulting in the end in the listing of his *On the Heavens* on the Index. The Holy Office possibly knew of the friendship between Galileo and Cremonini even though their ideas were very different. In fact, in 1604 their friendship was made clear when Galileo was accused before the Inquisition of Padua of practicing astrology. The accusation was considered of no importance by the Venetian government and it was dropped. But the inquisitor in Padua,

respecting the duties of his office, may have sent word of the accusation to the Holy Office, where there could have arisen the suspicion
that Galileo shared some of Cremonini's ideas and could, therefore,
have become involved in his trials.

From the minutes we know that Bellarmine was one of the seven
cardinals present at the meeting. Surely they spoke of the discoveries
and opinions of Galileo, which were causing such a fuss at Rome, and
expressed their concerns about them. But who was it that started the
search for whether Galileo was mentioned in Cremonini's trial? Some
have suggested that it was Bellarmine. Since the minutes of the Holy
Office report only the decisions taken and not the discussions, there is
no way to know with certainty. But one cannot doubt that the Jesuit cardinal took part in the discussions, and he must have been concerned, no
less than the others, about the theological consequences of Galileo's
discoveries. In Bruno's trial one of the most serious accusations was his
position that the universe was infinite and that there were an infinity
of inhabited worlds. Galileo's telescopic observations had now made
it clear that there were an enormous number of stars not visible to the
naked eye and this would probably have raised the specter of Bruno in
Bellarmine's mind. It was not by chance, it appears, that his first question to the Jesuit mathematicians at the Roman College concerned the
reality of "a multitude of fixed stars." That aside, Bellarmine must have
been worried about how scripture could be reconciled with Copernicanism, already clearly professed by Galileo. The problem must have
arisen frequently in the heated discussions that Galileo had with churchmen and with educated Romans. Bellarmine was, of course, at a wholly
different intellectual level than those who, like delle Colombe, opposed
Galileo insistently. Others with similar attitudes were surely not lacking at Rome. But the zeal with which Galileo very explicitly promoted
Copernicanism during his stay in Rome must have worried Bellarmine.
Surely, Galileo was a valued mathematician, but, as a layperson, he was
not competent in theology, and by this very fact was there not the danger that he would not take seriously enough the problems from scripture connected with Copernicanism? An indication of these concerns
of Bellarmine, as well as of other Church authorities, can be found in
the statements he made to the Tuscan ambassador, Guicciardini, at the

end of Galileo's stay in Rome: "If he [Galileo] had stayed here too much longer, they could not have failed to come to some judgment upon his affairs" (Galileo, *Opere,* 12:207).

The concern about Galileo certainly existed among the highest of Church authorities, but the severe secrecy of any proceedings of the Holy Office kept it from becoming public. Thus Galileo, having been warmly recommended by the grand duke of Tuscany, could receive public honors, at least as an astronomer. And so he was received in audience by Pope Paul V, who showed him great esteem. He received like esteem from many of the most prominent Church authorities, including Cardinal Maffeo Barberini, who later on, as Pope Urban VIII, would play a most important role in the Galileo Affair. Galileo was further recognized as an astronomer by his admission on April 25 as the sixth member of the Academy of the Lincei (Lynxs), one of the first scientific academies in Europe, having been founded eight years earlier by the young Prince Federico Cesi (1583–1630), who wished to promote the natural sciences. Thus a deep friendship was born between Galileo and Cesi, and it lasted until the premature death of the young prince. Galileo always felt himself highly honored as an academician and frequently signed himself as: "Galileo Galilei, Linceo." Another important event during Galileo's stay in Rome was the solemn academic assembly held in his honor at the Roman College, with the presence of many cardinals and important civil authorities of Rome. Odo Van Maelcote gave the official address, significantly entitled "The Starry Messenger of the Roman College." He first brought to the attention of the audience Galileo's telescopic discoveries and then reported on the observations made by the mathematicians at the Roman College which confirmed them. He prudently left to the listeners the conclusions to be drawn from the confirmed observations. It is strange that in the letters sent from Rome during this time Galileo never mentioned this academic assembly. Perhaps he had hoped that Clavius and his disciples would have taken a more explicit position in his regard. Some of the nuances in Maelcote's presentation almost certainly displeased him. For example, in quoting the long title of Galileo's *Starry Messenger,* Maelcote omitted Galileo's words describing the telescope as "discovered recently by him [Galileo]." Furthermore, Maelcote stated that the oval form of Saturn and the phases of Venus had been discovered by the Jesuit astronomers before they received word of Galileo's discoveries.

It would be wrong, however, to see in Maelcote's prudent presentation an indication of a coldness towards Galileo. It was Clavius's position that was reflected in Maelcote's prudence. By now Clavius was persuaded that the Ptolemaic system was no longer tenable. This is clear from what he said in the third volume of his *Mathematical Works,* which had just been published that year. After acknowledging the value of Galileo's discoveries with references to the *Starry Messenger,* he said: "Since that is the way things are, let the astronomers see how they can manage the celestial orbs, in such a way that they are able to save the phenomena." But these words do not mean that he had accepted Copernicanism. According to a statement by the Jesuit Athanasius Kircher (1602–1680) in 1633, Clavius and his Jesuit colleagues would not have been far from Copernicanism, but they had been "pushed and obliged to write in favor of the common opinions of Aristotle."

The original source of such an obligation goes back to the Constitutions of the Jesuits written by their founder, Ignatius of Loyola, which required fidelity to Aristotelianism in the teaching of philosophy. This obligation was repeated in the *Ratio Studiorum,* which, issued in 1599 by then Superior General of the Jesuits Claudio Acquaviva (1543–1615), established the norms for the studies of Jesuits and for their teaching. This insistence on Aristotelianism, it must be stressed, had nothing to do in the mind of Ignatius with Copernicanism. At the time of the writing of the Constitutions, *On the Revolutions* had been out for only about ten years and its philosophical implications had had little influence. The insistence on Aristotelianism by Jesuit superiors was based exclusively on their conviction that it formed the best basis for philosophy and, given the necessary modifications, for the so-called "preambles" to the Catholic faith, such as God's existence and the immortality of the soul.

Besides these general prescriptions was there any specific warning against Copernicanism by Jesuit superiors? As best as we can establish from existing documentation, up until the academic assembly at the Roman College there was none. Only a few days after that assembly, Acquaviva, still general of the Jesuits, sent a circular letter to all Jesuit professors in which he recommended "uniformity of doctrine," which implied Aristotelianism in philosophy, but also by implication in theology. No mention was made of Copernicanism.

Therefore, with or without specific norms, the cautious approach by Maelcote, Clavius, Grienberger, and other Jesuits in accepting Galileo's conclusions was dictated by the traditional obligation of faithfulness to Aristotle. And it was the ideal of Jesuit obedience, "in some ways blind" as Ignatius spoke in the Constitutions, that formed the roots of that obligation. In his famous *Letter on the Virtue of Obedience* Ignatius explained in detail that the highest form of obedience goes beyond the simple carrying out of a command or the will to do so, and requires the assent of the intellect. Certainly, adds Ignatius, the intellect is not free as the will is, and its assent is led spontaneously towards that which it sees as true. However, in certain cases, *when, that is, the intellect is not constrained by the evidence of the known truth,* it may tend, by reason of the will, in one direction or the other. *And when this case is presented,* whoever professes obedience should lean towards the will of the superior (*Epistula de virtute oboedientiae,* 1553; emphasis added).

As we can see from these words, that which was required of the Jesuits through obedience was not the sacrifice of intelligence *tout court* nor the abdication of intellectual responsibility. In the case where a Jesuit was faced with an *evident certainty,* no superior could demand that he assent to something contrary. The Jesuit astronomers of the Roman College themselves had shown that they were not slaves of an unconditionally blind obedience because at the end they had accepted Galileo's discoveries, motivated, as a matter of fact, by the "evidence of the known truth." And yet this implied, without a doubt, that they were thereby not faithful in a matter of great importance to the tenets of Aristotelianism. But here it was a question of a "known truth." Obedience could not force the Jesuits to deny what they had been able to confirm with their own eyes, such as the existence of Jupiter's satellites and the phases of Venus. But should they follow Galileo to the point of accepting Copernicanism, for which there were not yet any proofs? There was the system of Brahe, which the Jesuits knew quite well. So, for the moment at least, obedience held sway.

On the other hand, to accept Copernicanism meant denying, in addition to Ptolemy's system, a large part of the Aristotelian natural philosophy. But what natural philosophy could one call upon to substantiate the claim that the Sun and not the Earth must be at the center of the world system? It was also this knowledge of the physical problem that existed at the foundation of the new astronomical ideas that made the Jesuits hesitant.

Furthermore, there was the "theological" component and, to be more precise, the "scriptural" component. Certainly the Jesuits had also given consideration to the difficulties against heliocentrism deriving from scripture. These kinds of considerations were becoming always more important, as a result of the heated debates between Galileo and his antagonists during his stay in Rome. Undoubtedly, the influence of Bellarmine, the most authoritative Jesuit theologian, did not fail to have an impact on the prudent hesitation of the Jesuits. Bellarmine, as we know, was very troubled by the theological problems Copernicanism implied, and such concerns of his were inevitably reflected in the position of the Jesuit astronomers.

Given this situation should we not surmise that Clavius and the other Jesuit astronomers were taking a direction from that time on towards the astronomical compromise of Tycho Brahe? Maelcote had mentioned him in his discourse at the Roman College and had defined him as the "incomparable astronomer" (a further reason, no doubt, for Galileo's resentment). But the reference made was to his merits as a tireless observer, without any mention of his system. And yet it would have been quite natural to do so with respect to the phases of Venus, a phenomenon that could be explained in Tycho's system just as well as in that of Copernicus. As to Clavius, even though he held Tycho in great esteem as an observer, he had never shown up until the present any appreciation for his system. Clavius had never responded to a long letter from Tycho written at the beginning of 1600. On the contrary, in a letter to Magini he had remarked that Brahe was "confusing all of astronomy, because he wants to have Mars lower than the Sun." There is little doubt that, given the influence of Clavius on his brother Jesuits, they would also not have tended, at that time, towards the ideas of the Danish astronomer. In fact, from the correspondence of the circle of Clavius's disciples, the tendency towards the system of Copernicus as opposed to that of Tycho Brahe is quite clear. It would only be after the steps taken in 1616 and especially after the condemnation of Galileo in 1633 that the Jesuits, although reluctantly, would opt for the system of Brahe.

Galileo departed for Florence on June 4, 1611. He must have surely been happy with the results of his trip to Rome. Of course he had his share of opposition and disagreements to his open support of Copernicanism. But he had also seen signs of esteem and appreciation for his discoveries from important people in the Church and in scientific circles in Rome. The recognition he had received from Pope Paul V must have given him hope that Church authorities would not easily lend an ear to those who opposed him. No doubt, ignorant as he was of the growing concerns of the Holy Office (whose proceedings were protected by the strictest secrecy), Galileo gave too much weight to the public appreciation he received in high places. As a matter of fact, these concerns about his activity in favor of Copernicanism would only grow stronger as time went on.

Shortly after his return to Florence, Galileo found himself involved in a dispute that would further sharpen the contrast between him and the Aristotelians. This dispute began from a discussion that Galileo had with two professors of the University of Pisa with respect to ice floating on water. According to Aristotle's theory of condensation, ice was denser than water (because condensation is a property of coldness) and so the ice should sink in water. If it floated on water, that was due, according to the Pisan professors, to its wide, flat shape, which produced a resistance to submersion. Contrary to this statement, Galileo maintained instead, on the basis of the theory of Archimedes, that it was the greater or lesser density of a body with respect to water that caused it to sink or to float. Consequently, ice must be less dense than water. A few days later, Ludovico delle Colombe took up a position in the ranks of the Aristotelian professors by claiming that he was able to carry out experiments that would prove that it was the shape of a body which determined whether it would float or not on water. Thus was born the plan to hold a public debate between him and Galileo. But Galileo, at the suggestion of the grand duke, preferred to put in writing a resume of the discussions that had been held up until that time. On the following October 2 the dispute was taken up again on the occasion of a dinner given by the grand duke to honor Cardinals Ferdinando Gonzaga

(1587–1626) and Maffeo Barberini, who were passing through Florence at that time. During the debate Cardinal Gonzaga took the position of the defender of the Aristotelian point of view, Professor Flaminio Papazzoni (1572–1614), while Cardinal Barberini supported that of Galileo. The public debate ended with Galileo the victor and he decided to replace the manuscript that he had already sketched out with a true and proper treatise on hydrostatics. It was published in the spring of 1612 under the title *Discourse on Objects which Rest on Water or which Move in It,* and was dedicated to Cosimo II (Galileo, *Opere,* 4:49–150).

The *Discourse* was a great success, and so before the end of that year Galileo had a second edition printed with various additions and clarifications. But the work's success provoked not only letters of consent but also of fiery opposition. Among the opponents we should remember Ludovico delle Colombe, who towards the end of that same year composed a work dedicated to Giovanni de' Medici with the title *An Apologetic Discourse Concerning the Discourse of Galileo Galilei* (4:311–69).

So as not to make the contrasts more bitter, Galileo preferred to offer no direct response to these writings, but entrusted the task of responding to his friend, the Benedictine Castelli, who had in the meantime obtained the chair of mathematics at the University of Pisa. But, despite Galileo's prudence, this argument with delle Colombe and the other Aristotelians only served to create a deeper division between them and gave a further push to the formation in Florence of a group organized against Galileo. The members of the group would become known among Galileo's friends as the *colombi* (pigeons), an obvious allusion to the most fiery of their exponents (Ludovico delle Colombe and his brother, the Dominican friar Rafael).

Another dispute, under certain aspects even more serious because of its future developments, was the one in which Galileo became embroiled with the German Jesuit Christoph Scheiner (1573–1650) with respect to the priority of the "discovery" of sunspots. According to what Fulgenzio Micanzio would subsequently remember (1631) in a letter to Galileo, the latter would have shown sunspots for the first time to Paolo Sarpi (1552–1623) and himself at Padua in August 1610. But at that time Galileo made no mention in writing of these observations, perhaps because he had not attributed any special importance to them.

It seems that the first real observations, as Galileo himself stated, occurred towards the end of 1610 and the beginning of 1611. At any rate it is certain that he showed the sunspots during his stay in Rome to various Churchmen and others, among them Father Maelcote, who mentioned this in a letter to Kepler.

It seems that, before Galileo, sunspots were observed by the English astronomer Thomas Harriot (1560–1621), but his observations became known only in 1784 and published only in 1833. On the other hand, the first publication on sunspots was by a Dutchman, Johann Fabricius (1577–1613), and came out at Wittenberg in 1611. In that work Fabricius stated that the spots were neither clouds nor comets but that they belonged to the solar surface. By studying their motion Fabricius deduced the likelihood that the Sun was rotating on its own axis.

As to Scheiner, on the basis of what he himself states (Galileo, *Opere*, 5:25), his first observations of sunspots occurred in March–April 1611 together with his Jesuit assistant, Cysat. However, from the beginning not even Scheiner gave particular attention to the phenomenon. But his interest was suddenly awakened in October of that same year. It might be that the Jesuits in Rome had let him know about Galileo's observations, or that Scheiner had received word of the observations of Fabricius. In any case, Scheiner in November wrote a letter to Mark Welser (1558–1614) in Augusta providing him with the results of his observations. On December 12 and 26 Scheiner sent two other letters to Welser on his observations, and subsequently he published them, together with the first letter, under the title *Three Letters to Mark Welser on the Sunspots*. According to the directives of Jesuit superiors of those times regarding books that might be polemical and for which the Jesuit order did not wish to be responsible, Scheiner had chosen a pseudonym: "Apelles post tabulam latens" (Apelles hidden behind the painting). According to an ancient story Apelles, the great Greek painter of the fourth century BCE, used to hide behind his paintings at public exhibitions so that he could listen to the criticisms, knowing that as long as his presence was not known, they would be offered more freely. Scheiner obviously wanted it to be known that he would likewise gladly accept criticisms of his work in which he denied that the spots were part of the solar surface and proposed that they were small planets, like

Mercury and Venus. This agreed with the Aristotelian teaching of the incorruptibility of heavenly bodies.

In January 1612, Welser sent a copy of the *Three Letters* to Galileo with a request for his opinion on the matters raised. Because Galileo was in poor health and was not in Florence, he was unable to respond until four months later. In his long letter to Welser, Galileo criticized the reasons given by Apelles for his conclusion that the spots could not be on the body of the Sun. He noted that the spots did not keep the same spherical shape, but were produced and then dissolved, and did not appear to have a periodic movement about the Sun. He therefore concluded that they could not be minor planets as Apelles supposed.

Taken as a whole Galileo's response was prudent. Obviously, he was not yet sure of himself, given the newness and the difficulty of the material under study. But he nourished fond hopes that this new discovery would be of service to him "in tuning some reed in this great discordant organ of our philosophy." The allusion here is surely to the Aristotelian doctrine of the incorruptibility of the heavens, which would be proven false if, as he hoped, it could be conclusively shown that the sunspots were on the solar surface. Although he had had to contradict his conclusions, Galileo had shown himself to be very courteous towards Apelles, whose true identity he did not yet know, and he had not spared praising him.

Galileo had circulated his response to Welser to his friends. In particular, he had sent it to Cardinal Maffeo Barberini, together with the three letters of Apelles. And the cardinal had answered, praising him for the perspicacity of his genius and concluding that the opinion rejected by Galileo could not be true. But Galileo, having learned from previous experiences, wanted to arm himself against possible difficulties from scripture with respect to the sunspots. So he asked the opinion of Cardinal Carlo Conti about these matters. The cardinal answered in July 1612, stating that scripture did not support the Aristotelian theory of the incorruptibility of the heavens, but that, on the contrary, the common opinion of the Fathers of the Church was that the heavens were corruptible. As to the circular motion of the Earth, Conti stated that it was not very consistent with scripture and that, therefore, the Copernican hypothesis could be reconciled with scripture only if one held that the Bible spoke the ordinary language of the common people, an idea that, the cardinal added, "should not be admitted unless it is really necessary." Despite his personal caution, Conti did not hesitate to inform Galileo in complete honesty of the existence of Diego de Zuñiga's opinion in support of the possibility of a reconciliation between scripture and Copernicanism.

In the meantime Scheiner continued his observations and sent three new letters to Welser, which he then published in 1612, using his pseudonym and the title *A More Accurate Discourse*. Scheiner substantially repeated his theory that the sunspots were produced by small "stars" revolving about the Sun, among them at times was Venus, which according to him established the truth of the system of Tycho Brahe. This, of course, would not win a sympathetic response from Galileo. Other such statements of Scheiner about his priority of discoveries appear to indicate the beginning of a certain animosity of Scheiner towards Galileo.

In the dark as to the *A More Accurate Discourse*, Galileo had prepared a second letter addressed to Welser, which the latter received during the first days of October. Relying on numerous new observations, Galileo said that he was certain that the spots were not stars, but that they were close to the solar surface or at a distance from it which could not be detected. In addition to the growth and dissolution proper to each spot, they showed a motion in common in the course of which each kept its own latitude. From this Galileo deduced the rotation of the Sun on its own axis. Surely, he added, the fact that the spots belonged to the Sun's surface implied abandoning the Aristotelian theory of the incorruptibility of the celestial bodies. So, continued Galileo, Aristotle himself, if he were alive at the time, would have rather concluded in support of the corruptibility of the heavens.

Before he received the second letter of Galileo, Welser had sent him at the end of September a copy of the *A More Accurate Discourse* of Apelles. Thus, having just finished the second letter, Galileo felt obliged to send a third in order to answer the new publication of Apelles. The probability that the latter was a Jesuit was suggested a little later to Galileo by his friend Cigoli and was subsequently confirmed by Cesi. They both insisted that Galileo have his response to Welser published so as to prevent the "unnamed Jesuit" from usurping the priority of the discoveries. And thus, also through the intervention of Galileo's friends, the atmosphere was little by little being soured. This was already clear in the response of Galileo to Cesi. Alluding to the third letter of Welser, still being prepared, Galileo wrote: "but not for this should you [Cesi] be worried that much will be usurped because I hope to make it clear how foolishly this matter has been dealt with by the G. [*Gesuita,* Jesuit] to whom I wish to show such resentment as is fitting" (Galileo, *Opere,* 11:420).

This third letter of Galileo to Welser was finished on December 1 (5:186–249). In it Galileo states that he had read with interest *A More Accurate Discourse,* but he added that on various points he was not in agreement with the author. One of them was the refusal of Apelles to admit that the sunspots were part of the solar surface, by relying on the Aristotelian principle of the hardness and immutability of the Sun. On the contrary, Galileo showed, by a geometrical argument based on the very diagrams of Apelles, that the spots could not be located at any significant distance from the Sun. Galileo enunciated his hope that with these proofs Apelles would finally come to agree with his opinion and admit also the motion of the Sun on its own axis.

The three letters of Galileo to Welser were published in March 1613 under the editorship of the Lincean Academy with the title *A History and Some Demonstrations with Respect to the Sunspots* and contained a dedication to Galileo's great friend Filippo Salviati. The book was supplied with a preface written by the librarian of the Lincean Academy, Angelo de Filiis. In a polemical tone it claimed priority for Galileo's discoveries and, in particular, for the discovery of sunspots. When Galileo looked over it before it was printed, it must have made him uneasy with the foreboding that, by putting him in open argument with Scheiner, it ran the risk of damaging his relationship with the Jesuits, as one can deduce from the responses addressed to him by Cesi and Cigoli, who insisted that the claim to priority should be maintained, even though in a subtler tone than that of de Filiis's original text. But Galileo's foreboding was correct and this preface, despite the fact that it was toned down, provoked protests on the part of the Jesuits of the Roman College and later on claims of priority for Scheiner who must have been the most resentful of all. But he preferred for the moment to remain silent. On the other hand Galileo had never, in fact, in his three letters to Welser claimed priority for the discovery of sunspots.

As to the true identity of Apelles, Galileo only came to know of it in the following year, 1614. In that same year there had appeared a writing at Ingolstadt under the title *Mathematical Discourses on Astronomical Controversies and New Discoveries.* The title page listed a certain Locher Boius, said to be a disciple of Scheiner's. In this work, even though geocentrism was reasserted, Galileo's recent discoveries with the telescope were mentioned, with high praise for Galileo. The work added that, following such discoveries, Clavius and Magini had changed their system of the heavens. As to the sunspots, it showed a further evolution in Scheiner's thought that, with respect to many points, had become closer to Galileo's opinions.

In February 1615 Scheiner himself sent a copy of this little work to Galileo with the expectation that he would receive in turn some of the latter's publications. In the accompanying letter, the German Jesuit mentioned that he knew that Galileo tended towards Copernicanism, and he declared that he wanted to remain open to the dispassionate considerations of those who, like Galileo, reasoned differently than he did. Two months later he sent to Galileo another work of his entitled *Sol ellipticus* and asked for Galileo's opinion about it. Scheiner, however, did not get that opinion until ninenteen years later, in Galileo's *Dialogue.* And it would certainly not be the one he had hoped for.

Going back to Galileo's *Letters on Sunspots,* their publication required the imprimatur, that is, a permission to publish granted by Church authorities. In this case the censor required the elimination of the statement contained in the second letter to Welser whereby the incorruptibility of the heavens was said to be "not only false but erroneous and repugnant to those truths of Sacred Scripture about which there could be no doubt." Galileo had based this statement on the answer he had received on this matter from Cardinal Conti, but the censor had remained immovable in his request to abolish this reference as well as any other reference to sacred scripture. And so the passage had to be removed. On the other hand the censor had not objected at all to having a clear statement about Copernicanism in the third letter to Welser, probably because it appeared in a marginal note and, therefore, did not seem to constitute a true and proper profession of Copernicanism.

Perhaps Galileo did not take sufficient account of the importance of the censor's intervention with respect to the Bible. This intervention betrayed the preoccupation already existing in Rome since the time of Galileo's visit there, and even quite a bit before that, with respect to the new view of the world that Galileo had drawn from his telescopic observations. This preoccupation would only grow as the actions of Galileo and his followers in favor of Copernicanism slowly develop, driving not a few of the theologians to join up with the Aristotelian philosophers and the conservative astronomers.

THREE

The Scriptural Controversy Grows

A first indication that theologians were beginning to create close ties to the Aristotelians had occurred a little more than a year before in Florence. In the course of a conversation that had taken place among a group of Florentine intellectuals on November 1, 1612, the Dominican Niccolò Lorini (1544–?) had attacked the Copernican ideas as being contrary to scripture. Galileo sent a protest letter to Lorini, but this letter has been lost. But we have the response of Lorini himself written on November 5. The Dominican showed his surprise at being accused of discussing philosophical questions by anyone. But he admitted that he had made a reference, without any special involvement, to the "opinion of [that] Ipernicus, or whatever his name is," and he stated that it "appears to be against Holy Scripture." This answer must have made Galileo laugh, as he wrote to Cesi showing that he thought he had nothing to fear from an "incompetent conversationalist." But he was wrong.

A little more than a year later on December 12, 1613, at the court of the grand duke who was then at Pisa, something of much greater importance in the theological-biblical development of the Copernican controversy took place. Benedetto Castelli, who taught mathematics at the University of Pisa, gave Galileo the news in a letter written two days later. Castelli had taken part in a lunch offered by the grand duke together with Cosimo Boscaglia, a philosophy professor at the University of Pisa. Castelli, in response to a question of the grand duke, had spoken of the observations of the Medicean Planets, which he had carried out the night before. Boscaglia admitted that they were real, but he added that "only the motion of the Earth seemed incredible and could not be true, all the more so since Holy Scripture was clearly against this opinion." In the discussion that followed, Christina of Lorraine, the mother of the grand duke, appeared to take sides with Boscaglia, while the grand duke and the grand duchess, together with someone else, took sides with Castelli. The latter, speaking as a theologian, had blunted the biblical arguments of Boscaglia and had reduced him to silence.

Castelli's report was optimistic, but undoubtedly Galileo became very concerned about this scriptural development in the discussion. He saw that his adversaries, since they could no longer deny the reality of his discoveries, were proceeding ever more to dig themselves in behind the bastion of scripture. It was the last bulwark, but the one most to be feared, because it could nullify Galileo's plan to have Copernicanism accepted or at least considered without preconceptions in the Catholic world at that time.

It was all the more worrisome for Galileo that the discussion had taken place at the grand duke's court, even more so since the Grand Duchess Christina, a woman of strict piety, could be vulnerable to such propaganda against Copernicanism founded on scripture. Galileo, therefore, felt it necessary to establish a defense by making clear his position on the relationship between science and the Bible. He did this by writing a long letter to Castelli, which he certainly intended to have a wider circulation in manuscript form.

At the beginning of this letter Galileo admits that sacred scripture could not lie or deceive, but he immediately adds that its interpreters and expositors could err in various ways; the most serious error would be if they should wish to stop at the pure meaning of the words. In fact, in such a case, one would wind up attributing to God human forms and feelings such as anger, repentance, and hate. There exists, therefore, the possibility—even at times the necessity—of interpreting scripture in a nonliteral way. Such is the case, Galileo states, when it comes to arguments about natural phenomena. In fact, Galileo explains, both Holy Scripture (since it is dictated by the Holy Spirit) and nature (which faithfully carries out the divine orders) come from the divine Word. But while scripture has to be adapted to the common ability to understand and, therefore, has to use words and ways of speaking that, if taken in their literal sense, are far from the truth, nature, since it is "inexorable and unchangeable," is not at all concerned "that its recondite reasoning and ways of working be exposed to human abilities." Therefore, whatever "sense experience" puts before our eyes or whatever "necessary demonstrations" allow us to conclude should not be called in doubt on the basis of scriptural citations that, taken in the literal sense, would seem to say something different.

In other words, two truths that come from the same divine Word, the source of all truth, can never be in contradiction. Therefore, once we are sure (in the way described above) of certain "natural effects," it is the task of theologians to find the true sense of the scriptural passages that are related to those effects, so as to find agreement between the two truths.

Galileo is, therefore, warning against the danger of wishing to extend that which concerns articles of faith having to do with matters of eternal salvation to other matters that should be left open for free discussion. This danger is all the greater, he adds, when the extension is made by persons who "are clearly seen to be completely devoid of the information that would be required—I will not say to disprove, but—to understand the demonstrations with which the most acute sciences proceed in confirming some of their conclusions" (Galileo, *Opere*, 5:284). And to conclude he proposes his own interpretation of the passage from the Book of Joshua about the stopping of the Sun, which had been central to the discussions at the court of the grand guke. This passage had become by now a "classical" argument from scripture against the immobility of the Sun. Galileo went to all extremes to prove that, even accepting the literal interpretation, the passage was more in agreement with Copernicanism than with the geostatic theory.

Upon receiving this letter, Castelli had copies made that began to circulate. As noted already, Galileo himself had probably hoped this response of his to the objections based on scripture would made public.

But if he thought that this would have helped to calm the waters, he was grossly mistaken. In fact, the *Letter to Castelli* would only sharpen the tensions and strengthen the concerns of the theologians.

The tension that had been building finally exploded clamorously with a sermon that another Dominican, Tommaso Caccini (1574–1648), gave on December 21, 1614, in the Church of Santa Maria Novella in Florence. Caccini lived with Lorini in the Convent of St. Mark in Florence, and he was very much involved with the League of Pigeons. We know that he had read the severe criticism of Copernicanism in the manuscript of Tolosani, who had also lived in the same convent. Caccini took up this criticism and turned it on Galileo, against whom he had apparently also preached in Bologna in 1611.

Caccini's sermon was on chapter 10 of the Book of Joshua, which contained the passage about which Christina of Lorraine had questioned Castelli and with which Galileo dealt at the end of his *Letter to Castelli*. According to a version that circulated in that city quite a bit later, Caccini would have begun his sermon with the words in Latin from the book of the Acts of the Apostles (1:11): *Viri Galilaei, quid statis adspicientes in coelum?* (Men of Galilee, what are you looking for in the sky?), with an obvious allusion to Galileo and his followers, the *Galileisti*. Caccini had stated in his sermon that mathematics was a diabolic art and that mathematicians, as disseminators of heresies, should be driven from all of the states. Another Dominican, Luigi Maraffi, a great friend of Galileo, when he heard of the sermon wrote to him from Rome to show his regret that a member of his order had shown such "madness and ignorance."

Galileo would have liked to have gone right to the bottom of the affair by requesting atonement from Caccini. But when he asked Cesi's opinion, he was advised to abandon the idea because it would not have helped at all and could, on the contrary, have had serious consequences. In fact, Cesi added: "as to Copernicus's opinion, Bellarmine himself who is one of the heads in the congregation [of the Holy Office] concerning these matters has told me that he holds it to be heretical and that the motion of the Earth is without any doubt against scripture" (Galileo, *Opere*, 12:129).

Cesi obviously was afraid that with the controversy heating up a request might be made to the Holy Office for a response, which would run the risk that Copernicus's book would be prohibited. And so Galileo became convinced that it was better to remain quiet. But his opponents were not of the same mind. Lorini (who himself had deprecated the tone of the sermon of his religious companion Caccini) came a little later to possess a copy of the letter of Galileo to Castelli, and he discussed it with other Dominicans of the monastery of St. Mark in Florence. A little more than two years before, Lorini had shown that he evinced not great interest in "Ipernicus"; now the situation took a new course. In the *Letter to Castelli* Galileo had entered into theological matters and had pretended, even though he was only a simple layman, to deal with matters of biblical interpretation. That was extremely serious (the other fathers also agreed) because it set up an example of the kind of private interpretation of Holy Scripture that the Council of Trent had condemned.

In fact, the council had decreed in session 4 (April 8, 1546), decree 786:

> Furthermore, to control petulant spirits, the Council decrees that, in matters of faith and morals pertaining to the edification of Christian doctrine, no one, relying on his own judgment and distorting the Sacred Scriptures according to his own conceptions, shall dare to interpret them contrary to that sense which Holy Mother Church, to whom it belongs to judge of their true sense and meaning, has held and does hold, or even contrary to the unanimous agreement of the Fathers. (Trans. Blackwell, *Galileo, Bellarmine,* 11–12)

Certainly for Galileo the discussion of the Earth's motion lay outside the matters of faith, as he had clearly stated in his *Letter to Castelli.* But for his opponents it concerned matters of faith, or it was closely connected to such. The crux of the theological problem lay right there. Lorini, therefore, felt that he was obliged "in conscience" to alert the Roman ecclesiastical authorities to the matter. On February 7, 1615, he sent a "true copy" of the *Letter to Castelli* to Cardinal Paolo Sfondrati (1561–1618), prefect of the Congregation of the Index, so that it could be examined. The Congregation of the Index had been established by Pius V in 1571 with the purpose of preventing the distribution of printed material containing ideas contrary to Catholic faith and morals. Its activities were strictly linked to those of the Congregation of the Holy Office, which had been founded by Paul III in 1552 with the responsibility to control all matters concerning faith and morals.

In the accompanying letter, which Lorini had wished to be kept secret so that it would not be "taken as a court deposition," Galileo is not directly mentioned, but it speaks of the *Galileisti* as the authors of the *Letter to Castelli*. After saying that the letter supported the Copernican position, Lorini added that it contained propositions that all the fathers of the monastery of St. Mark considered to be "suspect or rash."

Was, in fact, this copy sent by Lorini a "faithful copy" of Galileo's original? All historians of science up to the present have denied that it is and have noted various discrepancies with respect to the presumed original published by Favaro (Galileo, *Opere,* 5:279–88), which they say were introduced so as to make the theological accusations against Galileo easier. However, it has recently been concluded that, in fact, the copy sent by Lorini was a faithful copy of the original *Letter to Castelli*. As Galileo himself stated he wrote this letter with a "running pen," that is, hastily, and did not, therefore, have time to refine certain statements, especially as regards scripture. On the other hand, Galileo must have had an inkling of the accusations of heresy in the *Letter to Castelli* brought by Lorini and other Dominicans of St. Mark's monastery as well as of the fact that Caccini had gone to Rome. This did not promise anything good, and I will address it in the next chapter. Galileo was obviously worried, and he revised the *Letter to Castelli* by softening some of the expressions and by giving a more cautious rendering to the various statements that touched upon theology. It was, in fact, such a revised copy that he sent on February 16 to his friend Monsignor Piero Dini with an accompanying letter in which he stated that it was:

> the correct version as I wrote it. I ask you to do me the favor of reading it along with the Jesuit Father Grienberger, a distinguished mathematician and a very good friend and patron of mine, and, if you deem it appropriate, of having it somehow come into the hands of the Most Illustrious Cardinal Bellarmine. The latter is the one whom these Dominican Fathers seem to want to rally around, with the hope of bringing about at least the condemnation of

Copernicus's book, opinion, and doctrine. (Galileo, *Opere,* 5:291; trans. Finocchiaro, *Galileo Affair,* 55)

As we see, Galileo was still counting on the Jesuits to help him. In fact, in a postscript to the letter to Dini he added: "I think the most immediate remedy would be to approach the Jesuit Fathers, as those whose knowledge is much above the common education of friars." Galileo hoped that this superior education of theirs would have driven them, if not to align themselves with him in favor of Copernicanism, at least to react against the intemperance of the ignoramuses.

The *Letter to Castelli* was not printed and, therefore, did not come under the jurisdiction of the Congregation of the Index. But it still concerned questions connected with the Catholic faith, and so Cardinal Sfondrati sent it, together with the accusatory letter of Lorini, to Cardinal Millini (1572–1629), secretary of the Holy Office. The consultor, appointed by Millini to examine the letter, had difficulties with only three of the statements in it, and they concerned precisely phrases that were present in Lorini's copy but not in the one sent by Galileo to Dini. The consultor commented, "They seem to sound bad when they are used with respect to Holy Scripture." But he added that two of the statements could be understood in the correct sense, and "For the rest, though it sometimes uses improper words, it does not diverge from the pathways of Catholic expression" (Galileo, *Opere,* 19:305; trans Finocchiaro, *Galileo Affair,* 136).

Although as a whole this opinion was not unfavorable to Galileo, it expressed some theological reservations. Therefore, the Holy Office wanted to go deeper into the matters. The Holy Office needed to have the original *Letter to Castelli,* in order to formulate a definitive judgment. And since the letter had been sent to Castelli, who was then living in Pisa, the request for the original was sent on to the archbishop and inquisitor of that city. Dini spoke of Galileo's fears to their common friend, Giovanni Ciampoli (1590–1643). Ciampoli was born in Florence and was ordained a priest in Rome in 1614. Galileo had known him at Florence in 1608. He made rapid progress in his ecclesiastical career and would be intimately involved in the Galileo Affair, especially at the time of the publication of the *Dialogue.*

Ciampoli wrote in turn to Galileo in a reassuring tone. But he informed him of advice from Cardinal Maffeo Barberini:

> [The Cardinal] told me only yesterday evening that with respect to these opinions he would like greater caution in not going beyond the arguments used by Ptolemy and Copernicus, and finally in not exceeding the bounds of physics and mathematics. For to explain the Scriptures is claimed by theologians as their field, and if new things are brought in even though to be admired for their ingenuity, not everyone has the dispassionate faculty of taking them just as they are said. One man amplifies, the next one alters, and what came from the author's own mouth becomes so transformed in spreading that he will no longer recognize it as his own. (Galileo, *Opere,* 12:146)

Dini had had many copies made of the *Letter to Castelli,* and he informed Galileo in a letter of March 7 that he had given a copy to Father Grienberger and one to Cardinal Bellarmine with whom he had spoken at length. According to Dini's report, the cardinal excluded the possibility that things would end with the prohibition of Copernicus, but that, in the worst case, notes should be added to *On the Revolutions* so as to present the doctrine as a pure mathematical expedient. And Bellarmine suggested that, for the sake of prudence, Galileo should speak in the same manner. According to the cardinal, against the immobility of the Sun there was especially the phrase from Psalm 18:6 (according to the Vulgate): the Sun "exulted like a giant which prepares to run its course." Dini had replied that this phrase could be interpreted as a common way of speaking, but the cardinal had said that: "this is not something to jump into, just as one ought not to jump hurriedly into condemning any one of these opinions." Dini added that Bellarmine had promised that he would speak of "these matters" with Father Grienberger.

Dini had judged, therefore, that it would be a good idea for him to visit Grienberger. The latter had said that he would have preferred that Galileo "first carry out his demonstrations and then get involved in discussing the Scripture." In regard to the arguments advanced by

Galileo in favor of Copernicanism, Dini added that "the said Father thinks they are more plausible than true, since he is worried about other passages of the Holy Writ."

Galileo answered Dini on March 23 with a second important letter in which he offered a retort to each point raised in the comments of Bellarmine and Grienberger. According to Galileo, Copernicus had spoken of the mobility of the Earth as a real fact and not as a mathematical hypothesis. Therefore, it was necessary to either accept or reject Copernicanism without any possibility of a compromise (which would occur, as a matter of fact, if one wished to accept it as a pure mathematical hypothesis). And Galileo added:

> Whether in reaching such a decision it is advisable to consider, ponder, and examine what he [Copernicus] writes is something that I have done my best to show in an essay of mine. . . . I have no other aim but the honor of the Holy Church and do not direct my small labors to any other goal. . . . Indeed, out of the same zeal, I am in the process of collecting all of Copernicus's reasons and making them clearly intelligible to many people, for in his works they are very difficult; and I am adding to them many more considerations, always based on celestial observations, on sensory experiences, and on the support of physical effects. (Galileo, *Opere*, 5:299–300; trans. Finocchiaro, *Galileo Affair*, 62)

The first writing to which Galileo here refers is the *Letter to the Grand Duchess Christina,* which Galileo was preparing precisely at that time. The second is most likely the *Discourse on the Ebb and Flow of the Sea,* which, after further elaborations and additions, would one day become the famous *Dialogue Concerning the Two Chief World Systems.* At the end of his letter, Galileo answers the scriptural argument, to which Bellarmine had alluded, by proposing an exegesis of Psalm 19:5–6: "he [the Sun] comes forth like a bridegroom from his canopy. He rejoices like a hero to run a race, going forth from the end of the heavens and his orbit to their ends, and nothing is hidden from his heat." This psalm was one of the most frequently quoted biblical passages in favor of the motion of the Sun. Galileo provided an interpretation of such

motion based on the discovered rotation of the Sun around its axis
that—according to a certain idea of "gravity" he proposes here—was
the cause of the motions of all the planets around it (including the
Earth, of course). Moreover, the Sun was also the source of all heat, in
accordance with the end of the psalm's text. Such an exegesis was certainly not done in a way as to win the cardinal's sympathy! And so Dini,
after having consulted Cesi, did not in the end give this letter to Bellarmine to read.

Without a doubt Galileo was encouraged in his Copernican campaign as well as in the composition of his *Letter to the Grand Duchess
Christina* by the publication of a work by the Carmelite theologian Antonio Foscarini entitled *Letter of the Reverend Father Master Antonio Foscarini, Carmelite, on the Opinion of the Pythagoreans and of Copernicus Concerning the Mobility of the Earth and the Stability of the Sun and the New Pythagorean
System of the World, etc.* It reproduced a letter sent by Foscarini himself
to the superior general of the Carmelites. On March 7 Cesi had sent a
copy to Galileo with an accompanying letter in which he commented:
"A work that certainly could not have appeared at a better time, unless
to increase the fury of our adversaries is damaging, which I do not believe" (Galileo, *Opere,* 12:150; trans. Drake, *Discoveries and Opinions,* 154).
As we shall see, the cautious optimism of that last comment would
prove to be quite wrong.

In his work Foscarini gave importance above all to the inadequacy
and unlikelihood of the system of Ptolemy. He then spoke of Galileo's
discoveries, thanks to which the Copernican hypothesis now appeared
to be more acceptable since it was simpler and fit the observations better. And in this regard he mentioned Clavius and his statement that it
was necessary to find a better system than that of Ptolemy. Foscarini
commented: "But what other system could one find better than that of
Copernicus?" Then he confronted the problem of the scriptural difficulties against the motion of the Earth. Foscarini maintained that, since
the truth is one, the truth of scripture could not be contrary to the
truth of the Copernican system, if one admitted that the latter could
be proven. It should, therefore, be possible, continued Foscarini, to reconcile with Copernicanism those scriptural passages that were causing
difficulty. Foscarini went on to consider those passages, which he gath-

ered into six classes, and he proposed six exegetical principles that, according to him, would remove the difficulties in question.

The Carmelite had come to Rome in March 1615 to preach for the season of Lent in a church of that city. Certainly he must have been informed that his work had been consigned to a consultor and that the judgment was severe. Foscarini had written a letter in Latin defending his opinion, and he sent a copy of it, together with a copy of his booklet, to Bellarmine in order to have an opinion about it from that authoritative cardinal. Bellarmine answered him on April 12, 1615, with the following letter:

My Very Reverend Father,

I have read with interest the letter in Italian and the essay in Latin which your Paternity sent to me; I thank you for one and for the other and confess that they are all full of intelligence and erudition. You ask for my opinion, and so I shall give it to you, but very briefly, since now you have little time for reading and I for writing.

First, I say that it seems to me that your Paternity and Mr. Galileo are proceeding prudently by limiting yourselves to speaking suppositionally and not absolutely, as I have always believed that Copernicus spoke. For there is no danger in saying that, by assuming the earth moves and the sun stands still, one saves all of the appearances better than by postulating eccentrics and epicycles; and that is sufficient for the mathematician. However, it is different to want to affirm that in reality the sun is at the center of the world and only turns on itself, without moving from east to west, and the earth is in the third heaven and revolves with great speed around the sun; this is a very dangerous thing, likely not only to irritate all scholastic philosophers and theologians, but also to harm the Holy Faith by rendering Holy Scripture false. For Your Paternity has well shown many ways of interpreting Holy Scripture, but has not applied them to particular cases; without a doubt you would have encountered very great difficulties if you had wanted to interpret all those passages you yourself cited.

Second, I say that, as you know, the Council [of Trent] prohibits interpreting Scripture against the common consensus of the Holy

Fathers; and if Your Paternity wants to read not only the Holy Fathers, but also the modern commentaries on Genesis, the Psalms, Ecclesiastes, and Joshua, you will find all agreeing in the literal interpretation that the sun is in heaven and turns around the earth with great speed, and that the earth is very far from heaven and sits motionless at the center of the world. Consider now, with your sense of prudence, whether the Church can tolerate giving Scripture a meaning contrary to the Holy Fathers and to all the Greek and Latin commentators. Nor can one answer that this is not a matter of faith, since if it is not a matter of faith "as regards the topic," it is a matter of faith "as regards the speaker"; and so it would be heretical to say that Abraham did not have two children and Jacob twelve, as well as to say that Christ was not born of a virgin, because both are said by the Holy Spirit through the mouth of the prophets and the apostles.

Third, I say that if there were a true demonstration that the sun is at the center of the world and the earth in the third heaven, and that the sun does not circle the earth but the earth circles the sun, then one would have to proceed with great care in explaining the Scriptures that appear contrary, and say rather that we do not understand them than that what is demonstrated is false. But I will not believe that there is such a demonstration, until it is shown me. Nor is it the same to demonstrate that by supposing the sun to be at the center and the earth in heaven one can save the appearances, and to demonstrate that in truth the sun is at the center and the earth in the heaven; for I believe the first demonstration may be available, but I have very great doubts about the second, and in case of doubt one must not abandon the Holy Scripture as interpreted by the Holy Fathers. I add that the one who wrote, "The Sun also riseth, and the Sun goeth down, and hasteth to his place where he arose," was Solomon, who not only spoke inspired by God, but was a man above all others wise and learned in the human sciences and in the knowledge of created things; he received all this wisdom from God; therefore it is not likely that he was affirming something that was contrary to truth already demonstrated or capable of being demonstrated. Now, suppose you say that Solomon speaks in accordance with appearances, since it seems to us that the sun moves (while the earth does so), just as to someone who moves away from the seashore on a ship it looks like the shore is moving. I shall answer that when someone moves away from the shore, although it appears to him that the shore is moving away from him, nevertheless he knows that it is an error and corrects it, seeing clearly that the ship moves and not the shore; but in regard to the sun and the earth, no wise man has any need to correct the error, since he clearly experiences that the Earth stands still and that the eye is not in error when it judges that the sun moves, as it also is not in error when it judges that the moon and the stars move. And this is enough for now.

With this I greet dearly Your Paternity, and I pray to God to grant you all your wishes.

At home, 12 April 1615.

To Your Very Reverend Paternity.

As a Brother,
Cardinal Bellarmine.
(Galileo, *Opere,* 12:171–72; trans. Finocchiaro, *Galileo Affair,* 67–69)

Even though Bellarmine's answer was a private one, considering the prestige of the cardinal in the world of theology of that time, it could be taken as an indication of the posture of the Church as it faced the Copernican problem.

There is a great deal of contrast among Galilean scholars in their judgment on this response of Bellarmine. Some praise it as a proof of the "truly scientific" mentality of the cardinal, who would thus have given a lesson in scientific methodology to Galileo. Others consider it to be a sign of the narrow-mindedness of the famous Jesuit. The fact is that Bellarmine was neither a positivist *ante litteram* nor an obscurantist by set purpose. Especially at the beginning of his teaching career at Louvain, he seems to be a man with a brilliant and argumentative intellect. But a great deal of his later life was spent in theological polemic with the Protestants, and this had provoked a progressive hardening of his positions, an instinctive suspicion towards anything that seemed to undermine the Catholic faith.

That is evident in the first point of his response to Foscarini. For Bellarmine there is no problem if Copernicanism is considered to be a pure mathematical hypothesis and, as such, perhaps superior to geocentrism. But it is a completely different matter if one considers the motion of the Earth to be real. For then not only does one set a challenge to the scholastic philosophers and theologians, but one goes against the faith by "rendering the Holy Scriptures false." This crucial statement is justified by Bellarmine's second point, which he develops in two stages. First of all the Council of Trent forbade that scripture could be interpreted against the common understanding of the Church Fathers. Now, goes Bellarmine's argument, the Fathers together with all modern exegetes agree on a literal interpretation of the biblical passages that speak of the motion of the Sun and the immobility of the Earth. Therefore, the Church cannot accept a scriptural exegesis to the contrary, and that is what would be required should Copernicanism be accepted as describing the real structure of the universe.

This reasoning of Bellarmine was weak on two accounts. First, there was no common consent of the Fathers as to the motion of the Earth because they had never explicitly considered it in common. In his *Letter to the Grand Duchess Christina,* Galileo will be correct in pointing out that the Fathers had followed the manner of thinking and speaking of everyone without ever having addressed the problem now involved in the discussion of Copernicanism. The second weak point was that the Council of Trent had spoken of the consent of the Fathers "in matters of faith and morals." Now both Foscarini and Galileo had insisted that the question of the motion of the Earth was not a matter of faith. Bellarmine knew this. But he responds by stating that, even if one admits that the biblical statement about the motion of the Sun and the immobility of the Earth is not a matter of faith on its own account *(ex parte objecti)*, it becomes a matter of faith by the fact of who speaks *(ex parte dicentis)*, namely, the Holy Spirit. Here we face a broadening of the sphere of matters concerning faith that many of the theologians of Bellarmine's time would not have accepted. If his exegetical principle is accepted, there is from a theological point of view no possibility of a future proof of Copernicanism. And so Bellarmine's response to Foscarini could well have ended here. The third point of Bellarmine's answer to Foscarini would appear, therefore, to show that

Bellarmine himself was not too sure of his principle. In fact, it is precisely the possibility of a future proof that is now considered. Galileo and Foscarini would certainly have subscribed to the beginning statement in Bellarmine's third point. But that would be the end of their agreement because, right after having admitted as a theoretical possibility a future proof of Copernicanism, Bellarmine confessed that he was "extremely doubtful" that such a demonstration could ever be given. The basis for such skepticism was again from scripture. Solomon was the wisest of men, his wisdom being a gift from God, and he had stated that "the Sun rises and sets and returns to its place." That means that he had affirmed the motion of the Sun. And to the objection that Solomon spoke according to appearances, Bellarmine responds with an argument from philosophy founded on the evidence of common experience: "because we clearly experience that the Earth stays still." If one accepts this two-fold argument, from both scripture and philosophy, the possibility of a future proof of Copernicanism is for all practical purposes excluded a priori.

To conclude, despite the kind tenor and moderate manner of expressing himself, Bellarmine had in his response clearly denied the ideas put forth by Foscarini and by Galileo himself with respect to possibility of reconciling Copernicanism with scripture. And some phrases betrayed a real concern. Foscarini was a theologian, and Bellarmine and other high Churchmen were profoundly opposed to his intervention in favor of Copernicanism and saw it as distinctly irresponsible in that time of confusion. Ciampoli had shown quite a bit more perspicacity than Cesi when a week later he wrote to Galileo: "because it treats of Scripture the book runs a serious risk of being suspended at the first Congregation of the Holy Office which is one month away" (Galileo, *Opere,* 12:160).

Bellarmine's response to Foscarini was also known to Galileo, who wrote a very detailed comment on the matter. As to the advice given indirectly to Galileo by Grienberger, that is, to furnish convincing scientific demonstrations first of all and then to enter into the scripture problems, Galileo commented in a letter to Dini in May 1615:

> To me the surest and swiftest way to prove that the position of
> Copernicus is not contrary to Scripture would be to give a host of

proofs that it is true and that the contrary cannot be maintained at all; thus, since no two truths can contradict one another, this and the Bible must be perfectly harmonious. But how can I do this, and not be merely wasting my time when those Peripatetics who must be convinced show themselves incapable of following even the simplest and easiest of arguments, while on the other hand they are seen to set great store in worthless propositions? (12:184)

In these last words there is perhaps a reference to Bellarmine and to that philosophical basis of his anti-Copernican conviction that the cardinal had expounded in his letter to Foscarini. But the real response to Bellarmine's thesis was given by Galileo in the letter addressed to the Grand Duchess Mother, Christina of Lorraine (Galileo, *Opere,* 5:309–48).

This new systematic and deeper treatment of the ideas contained in his letters to Castelli and Dini about the relationship of scientific research to Holy Scripture was probably completed in the summer of 1615. After beginning by making reference to his discoveries, Galileo emphasizes the attacks from the pulpit not only against him personally but against mathematicians in general. They wanted with these attacks to convince the public that the new astronomical ideas were heretical. In so doing, Galileo adds, his opponents showed how ignorant they were in thinking that the ideas expressed by him were in any way new. In fact, Copernicus was their author, and he in turn had taken them up again from ancient Greece. And, Galileo continues, Copernicus had published his theories in response to the request of and with the support of many prelates and had dedicated his book to Pope Paul III, without there having been "the slightest indication of a scruple on the part of the Church about his doctrine." How in the world, Galileo asks, could it be that this Copernican theory should be condemned right at this time when "we are discovering how well founded it is, based on manifest experience and necessary demonstrations?" For Galileo there is a simple explanation. The opponents have never read Copernicus's book, and they condemn the "new ideas" for reasons that are quite far from being scientific; in fact, they are based on a literal interpretation of those biblical passages where it is said that the Sun moves and the Earth stands still. Now since the scripture can never lie or err, they concluded that the theory that the Sun is motionless and the Earth moves is erroneous and damnable.

Of course, Galileo admits, the Bible can never err, but one must understand the true sense of what it says. At this point Galileo takes up what he had already said in the *Letter to Castelli* about biblical expressions that, if taken literally, would appear to be not only full of contradictions and far from the truth but also full of serious heresies and blasphemies. These expressions were used by the sacred writers in such a way as "to accommodate the capacities of the very unrefined and undisciplined masses." This is particularly true when scripture speaks of natural phenomena. It follows "that in disputes about natural phenomena one must begin not with the authority of scriptural passages but with sensory experience and necessary demonstrations." These latter, in fact, make clear the inexorability of natural events, which leave no doubt once it is established that one is dealing with facts and secure conclusions. And, because the truth is one, there can be no contradiction between the certain conclusions from this research into nature and the true meaning of scripture.

As we see, Galileo here formulates (he will develop it in the following parts of the *Letter to the Grand Duchess Christina*) that principle of the autonomy of the study of nature, which will become one of the hinges of modern scientific research. But, at the same time, he does not see this autonomy as opposed to the content of the Christian faith or without any relationship to it. On the contrary, it is precisely autonomous scientific research that will allow a better understanding of the obscure meaning of certain biblical passages concerning nature.

As a confirmation that scripture has no intention of entering into statements of a scientific character, Galileo notes how rare it is that the Bible makes astronomical references. After quoting two passages from St. Augustine's *De Genesi ad Litteram,* Galileo concludes by paraphrasing them:

Let us now come down from these matters to our particular point. We have seen that the Holy Spirit did not want to teach us whether heaven moves or stands still, nor whether its shape is spherical or like a discus or extended along a plane, nor whether the earth is located at its center or on one side. So it follows as a necessary consequence that the Holy Spirit also did not intend to teach us about other questions of the same kind and connected to those just mentioned in such a way that without knowing the truth about the former one cannot decide the latter, such as the question of the motion or rest of the Earth or the Sun. (Galileo, *Opere,* 5:319; trans. Finocchiaro, *Galileo Affair,* 95)

Galileo states that this conclusion coincides with what "an ecclesiastical person in a very eminent position [Cardinal Baronio]" had once said, namely, "that the intention of the Holy Spirit is to teach us how one goes to heaven and not how heaven goes." Galileo then concludes:

Because of this and because (as we said above) two truths cannot contradict one another, the task of a wise interpreter is to strive to fathom the true meaning of the sacred texts; this will undoubtedly agree with those physical conclusions of which we are already certain and sure through clear observations or necessary demonstrations. Indeed, besides saying (as we have) that in many places Scripture is open to interpretations far removed from the literal meaning of the words, we should add that we cannot assert with certainty that all interpreters speak with divine inspiration since if this were so then there would be no disagreement among them about the meaning of the same passages; therefore, I should think it would be very prudent not to allow anyone to commit and in a way to oblige scriptural passages to have to maintain the truth of any physical conclusions whose contrary could ever be proved to us by the senses and demonstrative and necessary reasons. (Galileo, *Opere,* 5:320; trans. Finocchiaro, *Galileo Affair,* 96)

As we see, in this passage, which is one of the most important in the letter, Galileo states the priority of scientific considerations over exegetical ones in cases where biblical passages deal with questions about nature. He distinguishes two cases. The first is where science has *already come* to secure conclusions. In this case it is up to the exegetes to discover the true sense of Holy Scripture that agrees with those conclusions. The second case is where there exists the *possibility* of a certain scientific conclusion in the future. Here exegetes should be very prudent and should avoid holding as true certain biblical interpretations that could be denied by the future scientific conclusions.

Having set up a foundation with these claims, Galileo goes on the counterattack. It is not he who is to be accused of wishing to interpret scripture according to his own whim and fancy, but on the contrary the accusation goes to many of his opponents, who use the Bible to condemn his discoveries. There remained the more delicate case of the theologians. With an obvious allusion to Bellarmine, Galileo declares that he has the highest regard for their "profound learning and their most holy lifestyle." But he confesses that he had some qualms about their claim "in disputes about natural phenomena . . . to force others by means of the authority of Scripture to follow the opinion they think is most in accordance with its statements, and at the same time they believe they are not obliged to answer observations and reasons to the contrary" (Galileo, *Opere,* 5:323–24; trans. Finocchiaro, *Galileo Affair,* 99).

The reason for this posture of theirs, claims Galileo, is that they think that theology, because it is the highest science, should be able to dictate the law also for the "inferior sciences." Even here, Galileo twists the argument around. Surely theology is the highest science, in the sense that its object is the highest possible and that its conclusions are founded on divine revelation. But it is precisely because it is such that "it does not come down to the lower and humbler speculations of the inferior sciences, but rather (as stated above) it does not bother with them, inasmuch as they are irrelevant to salvation." Therefore, he concludes: "officials and experts of theology should not arrogate to themselves the authority to issue decrees in the professions they neither exercise nor study" (Galileo, *Opere,* 5:325; trans. Finocchiaro, *Galileo Affair,* 100).

What then should be the attitude that a wise theologian would take? For his answer, Galileo turns back again to a text of St. Augustine:

There should be no doubt about the following: whenever the experts of this world can truly demonstrate something about natural phenomena, we should show it not to be contrary to our Scripture;

but whenever in their books they teach something contrary to the Holy Writ, we should without any doubt hold it to be most false and also show this by any means we can; and in this way we should keep the faith of our Lord . . . in order not to be seduced by the verbosity of false philosophy or frightened by the superstition of fake religion. (Galileo, *Letter to the Grand Duchess Christina,* quoting Augustine, *On the Literal Interpetation of Genesis,* 1.1, chap. 21; trans. Finocchiaro, *Galileo Affair,* 101)

In his comment Galileo takes up almost word for word a text of Pereira:

These words imply, I think, the following doctrine: in the learned books of worldly authors are contained some propositions about nature which are truly demonstrated and others which are simply taught; in regard to the former, the task of the wise theologians is to show that they are not contrary to Holy Scripture; as for the latter (which are taught but not demonstrated with necessity), if they contain anything contrary to the Holy Writ, then they must be considered indubitably false and must be demonstrated such by every possible means. So physical conclusions which have been truly demonstrated should not be given a lower place than scriptural passages, but rather one should clarify how such passages do not contradict those conclusions; therefore, before condemning a physical proposition, one must show that it is not conclusively demonstrated. Furthermore it is much more reasonable and natural that this be done not by those who hold it to be true, but by those who regard it as false. (Galileo, *Opere,* 5:327; trans. Finocchiaro, *Galileo Affair,* 101–2).

Various Galilean scholars have seen in this comment by Galileo on Augustine's text a patent contradiction of the fundamental thesis of the *Letter to the Grand Duchess Christina* on the autonomy of scientific research in matters to do with nature. In fact, in the case of "matters of nature taught but not necessarily demonstrated" he appears to give the final word to the theologians.

But I do not think that such a contradiction, in fact, exists, based first of all on the fundamental principle of hermeneutics whereby obscure passages of an author should be interpreted in the light of the clear passages. Now, this passage of the *Letter* is the only one where this contradiction seems to exist. Is it possible that Galileo was so easily prepared to come to a meeting of minds with his adversaries and thus compromise everything that he had already stated, as well as all that he will continue to uphold in the following parts of the *Letter*?

In the second place, Galileo does not at all give unreservedly the last word to the theologians in the case of "matters of nature taught but not necessarily demonstrated." He yields to them *only* if "there should be a matter contrary to the Holy Scriptures." And it seems to me that this "being contrary" must for Galileo be taken in the strictest sense, as it is in the text of St. Augustine, meaning opposed to a truth of the faith that is contained in Holy Scripture. In keeping with the principle that a truth in theology can never contradict a truth in science, if the first is certain (which would be the case for a truth of faith) any "scientific" statement opposed to it would be *definitively* false. Consequently, one could exclude a priori that a proof of the scientific statement could be found in the future.

Galileo, of course, was perfectly aware that the question between him and his opponents on theology was whether the immobility of the Earth and the mobility of the Sun was a matter of faith or not. For the theologians it was a matter of faith. For that reason, in the concluding passage Galileo could provide them with only this one proposal. Given that "scientific conclusions demonstrated to be true" could not really contradict scripture, theologians must see to it that, before condemning a scientific conclusion, such as the mobility of the Earth, the conclusion is indeed not demonstrated. This does not at all mean that Galileo authorizes them to condemn immediately once the absence of such a demonstration is shown. He is rather inviting them to be prudent. Before condemning a scientific proposition one must study and understand it. Galileo has no doubt, as the following passage shows, that such prudence and intellectual honesty would produce positive results. And, in the end, even without saying so, he is convinced that new arguments in the future in favor of Copernicanism will force theologians to a further review and consequent suspension of their judgment. And one fine day, Galileo is convinced, they will change that judgment.

And so, this much discussed passage of the *Letter* by no means appears to contradict the principles previously formulated by Galileo. And that seems to be confirmed by what he states with respect to the interpretation of scripture by the Fathers of the Church:

> From this and from other places [of St. Augustine, cited by Galileo previously] it seems to me, if I am not mistaken, the intention of the Holy Fathers is that in questions about natural phenomena which do not involve articles of faith one must first consider whether they are demonstrated with certainty or known by sensory experience, or *whether it is possible to have such knowledge and demonstration.* When one is in possession of this, since it too is a gift from God, one must apply it to the investigation of the true meanings of the Holy Writ at those places which seem to read differently. (Galileo, *Opere,* 5:322; trans. Finocchiaro, *Galileo Affair,* 105; emphasis added)

The discussion is now shifted by Galileo to the problem of the unanimous consent of the Church Fathers as to the motion of the Sun and the immobility of the Earth. As we know, Bellarmine, following the Council of Trent, had very much emphasized that this constituted, in fact, a rule of faith. Galileo responds that the common testimony of the Church Fathers can be validly applied "only to those conclusions which the Fathers discussed and inspected with great diligence and debated on both sides of the issue and for which they then all agreed to reject one side and to hold the other." But, states Galileo, the Fathers had never considered the Earth's motion and Sun's rest, given that in those times "this opinion was totally forgotten and far from academic dispute and was not examined, let alone followed, by anyone." Thus one cannot consider their witness as a rule of faith (Galileo, *Opere,* 5:335; trans. Finocchiaro, *Galileo Affair,* 108).

To confirm the fact that the hypothesis of the Earth's motion is not erroneous from the point of view of the faith, Galileo cites the opinion of Diego de Zuñiga in his commentary on the Book of Job (9:6), in which he "concludes the mobility of the Earth is not against the Scripture." He proceeds to deepen his argument with a clear allusion to the letter of Bellarmine to Foscarini:

> Furthermore, I would have doubts about the truth of this prescription, namely whether it is true that the Church obliges one to hold as articles of faith such conclusions about natural phenomena, which are characterized only by the unanimous interpretation of all the Fathers. I believe it may be that those who think in this manner may want to amplify the decrees of the Councils in favor of their own opinion. For I do not see that in this regard they prohibit anything but tampering, in ways contrary to the interpretation of the Holy Church or the collective consensus of the Fathers, with those propositions which are articles of faith or involve morals and pertain to edification according to Christian doctrine; so speaks the Fourth Session of the Council of Trent. (Galileo, *Opere*, 5:336–37; trans. Finocchiaro, *Galileo Affair*, 108–9)

But, Galileo emphasizes, the mobility or stability of the Earth or of the Sun are not questions of faith or of morals. As a confirmation of this, Galileo then quotes various other texts of Augustine, taken always from the *De Genesi ad Litteram*. These texts show "with what circumspection this most holy man walks before bringing himself to state resolutely any interpretation of Scripture as certain and so secure that one would have no fear of encountering some difficulty which would bring on some disturbance" (Galileo, *Opere*, 5:339).

At the end Galileo confronts the most important and at the same time most delicate question of the authority that the Church possesses to make decisions with respect to questions having to do with nature that are related to sacred scripture. He emphasizes first of all that "it belongs to no one other than the Supreme Pontiff or the Holy Councils to declare a proposition erroneous." Certainly no pope nor council had ever prohibited the Copernican theory, and this confirmed for him that geocentrism was not, at least directly, *of the faith*. But it is just as certain that the pope has the authority to make a decision now on Copernicanism and eventually even to condemn it. It is precisely this that caused Galileo the deepest worries and which constituted the final motive for this *Letter to the Grand Duchess Christina*. Therefore, the words that Galileo writes on this matter constitute altogether the most important conclusion of this writing and an indirect but firm warning directed to Pope Paul V himself:

> For in regard to these [Copernican] and other similar propositions which do not directly involve the faith, no one can doubt that the Supreme Pontiff always has the absolute power of permitting or condemning them; however, no creature has the power of making them be true or false, contrary to what they happen to be by nature and de facto. So it seems more advisable to first become sure about the necessary and immutable truth of the matter, over which no one has control, than to condemn one side when such certainty is lacking; this would imply a loss of freedom of decision and choice insofar as it would give necessity to things which are presently indifferent, free, and dependent on the will of supreme authority. In short, if it is inconceivable that a proposition should be declared heretical when one thinks it may be true, it should be futile for someone to try to bring about the condemnation of the earth's motion and the sun's rest unless he first shows it to be impossible and false. (Galileo, *Opere,* 5:343; trans. Finocchiaro, *Galileo Affair,* 114)

The strictly doctrinal part of the *Letter to the Grand Duchess Christina* ends here. Following it as a kind of appendix is an exegesis of the passage from Joshua 10:12–14 which presents a reworking of what Galileo had already placed at the end of the *Letter to Castelli.*

Substantially Galileo reasons that the miracle whereby the Sun stood still "could not in any way have occurred, should the heavenly motions be determined by the Ptolemaic system," whereas with the "Copernican system one can call upon the most literal and easiest interpretation." This attempt at a literal interpretation of the Bible using Copernicanism has also been severely criticized by many scholars as an open contradiction of the fundamental thesis of the *Letter,* namely, that the scriptures were not written to teach us "how the heavens go."

Although not wishing in any way to defend the contents of this "concordist" attempt of Galileo, I think it is important to try to understand the significance of it, that of an ad hominem argument. The biblical passage in question, perhaps more than any other, had been quoted in support of the Sun's motion and, therefore, of Ptolemaic geocentrism. Galileo puts himself on the same plane of his adversaries to show that, even if one persists in a literal interpretation of the passage, this is practically impossible in the Ptolemaic hypothesis, while it becomes possible and simple in the Copernican hypothesis. As any ad hominem argument, this is only a tactical argument for the sake of controversy, and not heuristic or directed at the advancement of knowledge.

Galileo had certainly given great importance to the *Letter to the Grand Duchess Christina,* having dedicated a great deal of time composing it. But in the meantime the situation had become increasingly more tense, with the result that Galileo and his friends decided not to have the *Letter* circulate, at least for the moment, except among a small circle of trusted friends. And so for the time being the *Letter* had little or no influence. As we shall see, it came to be known only some fifteen years later on the eve of Galileo's trial, and its real circulation will come still later after its printing in Strasbourg in 1636.

While Galileo was writing this *Letter to the Grand Duchess Christina,* new developments were happening in Rome in the way of activity against Galileo. On the one hand, it is true that the activity of Lorini did not develop as the latter had hoped. The Holy Office had given to the archbishop of Pisa, where Castelli lived, the task of obtaining the original of Galileo's *Letter to Castelli.* After much delay Galileo had finally returned the *Letter to Castelli* to the latter, but with explicit condition that he would not deliver it to the archbishop of Pisa as the latter had requested, but only read it to him. Following the instructions of the Holy Office, the archbishop was careful not to let Castelli know the reason for the request to have an original of the *Letter.* But Galileo must have deduced the real reason. On the other hand, if the copy of the *Letter* delivered by Dini to Bellarmine as the "true copy" was actually the one corrected by Galileo and, therefore, different from the original, to give now the archbishop the real original ran the risk that Galileo could be accused of falsehood. Galileo's advice to Castelli not to hand over the original but only to read it (which in fact Castelli did) seems to confirm the fact that Galileo had personally corrected the original in order to avoid the criticism of certain statements made therein.

In the meantime the Holy Office had to busy itself with another accusation against Galileo and the *Galileisti,* this time brought on by Tommaso Caccini. He had gone to Rome towards the end of February, both to obtain a coveted office at the Dominican convent of the Minerva and in order to bring to completion the attack that Lorini had begun. To that end he had a meeting with Cardinal Agostino Galamini, who was also a Dominican as well as a member of the Holy Office. In the meeting of March 19, Galamini let it be known that Caccini, convinced of Galileo's errors, wished to make a deposition concerning them, "for conscience sake." The deposition took place on the following day, in the presence of the commissary general of the Holy Office, the Dominican Michelangelo Segizzi. In it Caccini stated that Father Ferdinando Cimenes, of the Church of Santa Maria Novella in Florence, had heard from some of "Galileo's disciples" "scandalous propositions about God and the saints." Caccini then went on to a direct attack on Galileo, and stated that in the copy of the *Letter to Castelli* shown to him by Father Lorini, it seemed to him that "there is contained with respect to theological matters a doctrine which is not good." And he concluded:

> Thus I declare to the Holy Office that it is a widespread opinion
> that the above-mentioned Galilei holds these two propositions:
> the earth moves as a whole as well as with diurnal motion; the sun
> is motionless. These are propositions which, according to my con-
> science and understanding, are repugnant to the divine Scripture
> expounded by the Holy Fathers and consequently to the faith, which
> teaches that we must believe as true what is contained in Scripture.
> And for now I have nothing else to say. (Galileo, *Opere,* 19:308–9;
> trans. Finocchiaro, *Galileo Affair,* 138)

During the questioning period that followed his deposition, Caccini gave importance to the friendship of Galileo with the Servite Paolo Sarpi, "so famous in Venice for his impieties." Ten years before, Sarpi had defended the rights of the Venetian Republic in a dispute between the latter and the pope, which ultimately led to the interdict decreed by Paul V against the republic. He was therefore a quite unpopular person in Rome. And friendship with him was an aggravating circumstance against Galileo. Caccini asserted also that both Father Lorini and Father Cimenes held Galileo as "suspect in matters of faith," because of his opinions on the stability of the Sun and the mobility of the Earth, and for his pretension at interpreting scripture against the common opinion of the Holy Fathers. And he added:

> This man, together with others, belongs to an Academy—I do not know whether they organized it themselves—which has the title of "Lincean." And they correspond with others in Germany, at least Galileo does, as one sees from that book of his on the sunspots. (Galileo, *Opere,* 19:310; trans. Finocchiaro, *Galileo Affair,* 140)

Germany was the homeland of Protestantism, and Galileo's relations with Germany were emphasized by Caccini in order to discredit Galileo. As a matter of fact, the relations with Germany that resulted from Galileo's book on sunspots were those with Welser and with the Jesuit Scheiner, both of them certainly not Protestant. It is possible, however, that Caccini did not know the true identity of Scheiner.

The investigation by the Holy Office to ascertain the truth of Caccini's accusations was drawn out over a long period of time. At the end, the only point ascertained was Galileo's Copernican convictions. But it was a point of sufficient concern, the more so given the confusion created among educated people by Foscarini's letter.

Since Galileo's letters on sunspots had been mentioned in Caccini's deposition, on November 25 the Holy Office decided that the work should be examined. The opinion given by the censor on the matter has not been preserved, but it must not have been adverse to Galileo because this work of his would not be touched by the dispositions of the Congregation of the Index in the following year. Nevertheless, the fact remained that there was too much evidence with respect to Galileo's Copernican convictions to permit any further doubt about it.

But let us return to Galileo. Given the extreme secrecy that surrounded the proceedings of the Holy Office, he (as well as his Roman friends) had been left in the dark about the content of the actions of Lorini and Caccini. He became, however, increasingly concerned that a condemnation of Copernicanism by the Church could occur. The *Letter to the Grand Duchess Christina* had been an attempt at facing that danger with a written defense. But the increase of the tensions and suspicions against him had dissuaded Galileo from having it circulated, so that attempt had not brought about any concrete result. On the other hand, the Aristotelians as well as some friars like Lorini and Caccini were carrying on a systematic campaign of disparaging Galileo, which risked making an ever deeper impression upon the ecclesiastical authorities. To avoid their having the upper hand, Galileo judged it necessary to take action with a plan of oral defense, which obviously had to be carried out in Rome.

Galileo's Roman friends tried to dissuade him. Dini advised him rather "to be quiet and to strengthen his position with good and well-founded reasons both from Scripture and from mathematics and at the right time to put them forth with greater satisfaction." Galileo, however, was not persuaded, and at the end of autumn he decided to depart for Rome. Having been informed about it by the Grand Duke Cosimo II, the ambassador of Tuscany in Rome, Pietro Guicciardini, showed his apprehension concerning this initiative of Galileo, with words of cold realism. "I know, that certain friars of St. Dominic, who are in the Holy Office, and others are ill disposed toward him and this is no fit place to argue about the Moon, or, especially in these times, to try to bring in new ideas" (Galileo, *Opere,* 12:207; trans. Santillana, *The Crime of Galileo,* 110). This was, however, a down-to-earth approach that obviously Galileo could not share.

FOUR

The Copernican Doctrine
Is Declared to Be Contrary
to Holy Scripture

In spite of Ambassador Guicciardini's contrary opinion, Galileo put in practice his plan to go to Rome, where he arrived in the first days of December 1615. Judging from the first conversations upon his arrival, he wrote an optimistic letter to Secretary of the Grand Duke Picchena. Later on, however, he became aware of the "very vigorous impressions" made by his opponents on many authoritative persons, and of the necessity of taking an action in depth to neutralize them.

Galileo began, therefore, to carry out a program of feverish activity, availing himself of all the resources of his very skillful polemical art. Having been silenced with respect to science, and often ridiculed by him, Galileo's opponents took revenge by spreading malicious rumors and calumnies about him. And this forced him to prolong his stay in Rome, in order to destroy them. As time went on, the situation became ever more confused and tense. And even Galileo's friends had to employ many precautions in dealing with him so as not to raise suspicions and criticisms.

Among Galileo's opponents in Roman educated circles, a special place was held by the priest Francesco Ingoli (1578–1649), who had probably known Galileo in Padua. Ingoli was at the service of an influential member of the Congregation of the Index, Cardinal Bonifacio Caetani, and was a person of wide interests, among them astronomy. Always taking an active part in theological, philosophical, and scientific discussions, he had one such discussion with Galileo in the house of Lorenzo Magalotti, a future cardinal. He then put in writing directed to Galileo his defense of his anti-Copernican position, under the title *Disputation on the Location and Stability of the Earth*. Appointed in March of the following year as consultor of the Congregation of the Index, Ingoli would extend his anti-Copernican activity even with respect to Kepler. Galileo was prevented from answering to Ingoli by the decisions taken by the Church in February of the following year, and he would not be able to do so until eight years later.

At the beginning of February, Galileo had a meeting with Caccini at the latter's request. According to the report Galileo gave to Picchena, the Dominican had excused himself for the words used in his sermon and had offered to give to Galileo whatever satisfaction he might desire. And he had tried to make him believe that he was not responsible for the rumors that had spread about Rome concerning Galileo. When other visitors joined them, the conversation shifted to the Copernican controversy and as the conversation proceeded Caccini showed himself to be "very far from understanding what would be required in these matters." At the end, Galileo wrote, Caccini "went back to his first reasoning and sought to dissuade me from that which I know for certain" (Galileo, *Opere,* 12:231).

About the same time, Galileo had asked the grand duke for permission to go to Naples, undoubtedly so that he could meet with Father Foscarini. The intent must have been to specify with the Carmelite a decisive effort in the Copernican campaign, so much more so since Galileo was by this time convinced that he had an important argument in support of the Copernican system: the one derived from the phenomenon of the tides, an idea that had first occurred to Galileo about twenty years earlier. It was put in the form of a letter with the title *Discourse on the Ebb and Flow of the Tides,* and had been sent a month earlier to the very young Cardinal Alessandro Orsini (1593–1626). The latter was an admirer of Galileo and one in whose help Galileo had placed great hopes, as he himself declared to Picchena on February 6, asking for a special recommendation of the grand duke to the cardinal. Orsini, as we will soon see, took to heart Galileo's problem, but with quite a different result than the one Galileo had hoped for.

In spite of the situation, Galileo had not yet completely lost his optimism. He had the impression that the doubts concerning his orthodoxy had by now been cleared up to the satisfaction of the Roman authorities. And this must have, at least in part, reassured him. He was however totally in the dark with regard to the decisions that were maturing at that moment against Copernicanism. The Church authorities were by now too much concerned about the repercussions created by Galileo's pro-Copernican activities and by the *Letter of Foscarini* to be able to hesitate any longer in taking an official decision on the matter.

In fact, on February 19, two propositions on the Copernican system were submitted for the examination of the "qualifiers," the theologian consultors of the Holy Office, whose duty was to give the theological qualification of the said propositions. They were thus formulated:

1) The sun is the center of the world and completely devoid of local motion. . . .
2) The earth is not the center of the world, nor motionless, but it moves as a whole of itself, and also with diurnal motion.
(Galileo, *Opere,* 19:320; trans. Finocchiaro, *Galileo Affair,* 146)

These two propositions had been taken almost word for word from Caccini's deposition.

Among the qualifiers to whom the examination had been entrusted were well-known theologians, most of them Dominicans. None of them was competent in the field of astronomy, and yet they had no fear about giving an answer, and in the short period of time (less than four days) granted to them. Obviously, in the unshakable certainty of their philosophical and theological convictions, the qualifiers did not consider it necessary that they have more time in order to pass their judgment. And, as for that, after months in Rome of heated arguments about the case, they must have already had their minds clearly made up on the matter. Without a doubt, their opinion was agreed upon in the qualification meeting of February 23. On the following day there was the plenary session of the qualifiers and the other consultors of the Holy Office to set forth the definitive formulation of the qualifications to be given to the two propositions. On the first:

> All said that this proposition is foolish and absurd in philosophy, and formally heretical, since it explicitly contradicts in many places the sense of Holy Scripture, according to the literal meaning of the words and according to the common interpretation of the Holy Fathers and the doctors of theology. (Galileo, *Opere,* 19:321; trans. Finocchiaro, *Galileo Affair,* 146)

On the second:

> All said that the proposition receives the same judgment [qualification] in philosophy and that in regard to theological truth it is at least erroneous in faith. (Ibid.)

It is to be remarked that in theological terminology a proposition is "formally heretical" if it is judged directly contrary to a doctrine of faith—it is the most serious theological censure possible. A proposition "erroneous in faith" is not directly opposed to scripture but becomes such because it is a necessary conclusion drawn from a formally heretical proposition on which it depends. In the case of the Earth, its immobility could not be proved with certainty on the basis of biblical passages. Therefore, its negation was not formally heretical. However, its negation in the Copernican system was a necessary consequence of the (heretical) affirmation of the Sun's immobility. Such a negation was, therefore, at least erroneous in faith.

On the same day (Wednesday, February 24), a consistory of cardinals convened in the presence of Paul V. During the meeting, Cardinal Orsini pleaded Galileo's cause. It could not have been a less propitious moment. According to the report sent eight days later by Ambassador Guicciardini to Picchena:

The pope told him it would be well if he persuaded him [Galileo] to give up that opinion. Thereupon Orsini replied something, urging the cause, and the pope cut him short and told him he would refer the business to the Holy Office. As soon as Orsini had left, His Holiness summoned Bellarmine, and after discussing the matter, they decided that the opinion was erroneous and heretical; the day before yesterday, I heard they had a congregation on the matter to have it declared such. (Galileo, *Opere,* 12:24)

We do not know from whom Guicciardini would have had this information. The consistory's secret was not too strict, and therefore the Orsini intervention and the pope's response had come easily to the ambassador's knowledge. The same holds true for Paul V's summons of Bellarmine. As to the meeting of the Holy Office, Guicciardini had made a mistake. "The day before yesterday" would have been March 2, but the meeting took place, as we will see, on March 3. For the rest, Guicciardini could do nothing but make conjectures, given also the strictest secrecy that surrounded the procedures of the Holy Office. And from the documents we know instead that the qualification of the Sun's immobility as a heresy had been already given in the day previous to the consistory by the qualifiers and had been confirmed on that same day of the consistory during the plenary session of all the consultors. It was therefore not "decided" by Paul V and Bellarmine, on their own initiative. This is confirmed by the minutes of the weekly meeting of the cardinals in the presence of the pope, which was held on the day following the consistory, that is, on February 25. Undoubtedly, in the first part of the meeting, the assessor of the Holy Office and the commissary informed the pope and the cardinals on the censures that had been approved on the previous day, together with the questions connected with the issue of the Copernican theory (the "Galileo problem" included). As customary, these two officials of the Holy Office left the room and the second part of the meeting started, secret and reserved to the pope and the cardinal inquisitors. This is why the minutes, written during the third part of the meeting after the assessor and the commissary had come back, make no mention of that second part. And this explains also the necessity of informing those two officials as to what the pope and the cardinals had decided during that same second part of the meeting, as is clear from the following document:

> The Most Illustrious Lord Cardinal Millini notified the Reverend Fathers Lord Assessor and Lord Commissary of the Holy Office that, after the reporting of the judgment by the Father Theologians against the propositions of the mathematician Galileo (to the effect that the sun stands still at the center of the world and the earth moves even with the diurnal motion), His Holiness ordered the Most Illustrious Lord Cardinal Bellarmine to call Galileo before himself and warn him to abandon these opinions; and if he should refuse to obey, the Father Commissary, in the presence of a notary and witnesses, is to issue him an injunction to abstain completely from teaching or defending this doctrine and opinion or from discussing it; and further, if he should not acquiesce, he is to be imprisoned. (Galileo, *Opere*, 19:321; trans. Finocchiaro, *Galileo Affair*, 147)

As a matter of fact, it is probable that the censure of the qualifiers had already been privately communicated to Paul V by Bellarmine himself on the previous day, on the occasion of his encounter with the pope immediately after the consistory of cardinals. Bellarmine could in fact have had the time to be informed about the censure, which had been approved in the plenary meeting of all the consultors of the Holy Office on that same day.

In light of the attached censures, the Earth's motion defended by Galileo was at least erroneous in faith. And the Sun's immobility was even a heresy. But how to proceed with Galileo personally? He was by now famous throughout Europe and the "Mathematician and first Philosopher" of the grand duke of Tuscany. Furthermore, there was no doubt about the sincerity of his faith, despite his astronomical ideas. It was probably Bellarmine himself who proposed to Paul V, on the occasion of that private meeting with him, the procedure of a private warning. And this would explain the fact that the task to present such a warning was given by the pope precisely to him. With this expedient, Galileo would be silenced once and for all, without offending the grand duke. The eventuality that Galileo would dare to refuse to submit must have seemed extremely remote to Paul V and to Bellarmine. And this was what officially was decided during the secret part of the meeting of February 25. On the other hand, from the legal point of view, one had to foresee all the possibilities, as remote as they could have been. And this explains the presence of the two other further, possible phases of the procedure concerning Galileo.

As to Copernicanism and the theologians who supported it, the Congregation of the Index (of which Bellarmine was also a member) would see to neutralizing both of them in a convenient way. Bellarmine carried out the task assigned to him on the following day. We know it from two further documents, contained in the same file of the *priocessi* (trial) of Galileo, kept at present in the Secret Vatican Archive. The first chronologically comes immediately after that of the session of February 25:

> Friday, the 26th of the same month.
>
> At the palace of the usual residence of the said Most Illustrious Lord Cardinal Bellarmine and in the chambers of His Most Illustrious Lordship, and fully in the presence of the Reverend Father Michelangelo Segizzi of Lodi, O.P. and Commissary General of the Holy Office, having summoned the above-mentioned Galileo before himself, the same Most Illustrious Lord Cardinal warned Galileo that the above-mentioned opinion was erroneous and that he should abandon it; and thereafter, indeed immediately (*successive et incontinenti*), before me and witnesses, the Most Illustrious Lord Cardinal himself being also present still, the aforesaid Father Commissary, in the name of His Holiness the Pope and the whole Congregation of the Holy Office, ordered and enjoined the said Galileo, who was himself still present, to abandon completely the above-mentioned opinion that the sun stands still at the center of the world and the earth moves, and henceforth not to hold, teach, or defend it in any way whatever, either orally or in writing; otherwise the Holy Office would start proceedings against him. The same Galileo acquiesced in this injunction and promised to obey.
>
> Done in Rome at the place mentioned above, in the presence, as witnesses, of the Reverend Badino Nores of Nicosia in the kingdom

of Cyprus and Agostino Mongardo from the Abbey of Rose in the diocese of Montepulciano, both belonging to the household of the said Most Illustrious Lord Cardinal. (Galileo, *Opere,* 19:321–22; trans. Finocchiaro, *Galileo Affair,* 147–48)

The second document was written later, and it is found at the beginning of the minutes of the Holy Office's meeting of the following March 3, in which notice was taken of the decree of the Congregation of the Index against the Copernican writings (of which I will speak shortly):

> The Most Illustrious Lord Cardinal Bellarmine having given the report that the mathematician Galileo Galilei had acquiesced when warned of the order by the Holy Congregation to abandon the opinion which he held till then, to the effect that the sun stands still at the center of the spheres but the earth is in motion. (Galileo, *Opere,* 19:278; trans. Finocchiaro, *Galileo Affair,* 148)

The difference between these two documents is obvious. The second speaks only of a "warning" and of an "acquiescence," without any mention of Segizzi's intervention. Instead, in the first document that intervention of the commissary, *successive et incontinenti,* that is, immediately after Bellarmine's admonition, is put in particular evidence. But there is no statement that such an intervention had been motivated by the refusal of Galileo to accept the admonition of the cardinal, as prescribed by the instructions. Moreover, the first document lacks signatures of Galileo, of Segizzi, of the notary who wrote it, and of the witnesses named therein.

Many hypotheses have been put forward, since the second half of the nineteenth century, to explain the discrepancies between these two documents. One has held that the second part of the first document (that is, the one with the Segizzi injunction) is a fraud, added to the first part of it during the investigatory phase of Galileo's trial, in 1632, in order to be able to bring accusations against him at the Holy Office. In favor of such a hypothesis there is the fact that this document happens to have been "discovered" right at that time in the archives of the Holy Office. Others, on the contrary, have advanced the hypothesis that

Commissary Segizzi, disillusioned by the too moderate manners in which Bellarmine had delivered the admonition and by the prompt assent of Galileo, decided to omit the official report of this, and had inserted in the minutes the part containing his injunction, which in fact never took place but which he had hoped to be able to deliver.

Nevertheless, on the basis of recent research on the archives of the Holy Office, as well as of a calligraphic expertise, the hypothesis of a fraud perpetrated on the eve of Galileo's trial, in 1632, appears now to be indefensible. The calligraphic analysis of the document, obtained by me in the year 2009 from a great specialist of manuscripts of Galileo's time, Isabella Truci, of the National Central Library of Florence, indicates beyond a doubt that its entire recording was made by the same person, who has been identified as the notary of the Holy Office, Andrea Pettini, who wrote all the other documents of 1616. And such recording was made, as was customary, in the form of *imbreviatura,* that is, in an abridged form. As such, the signatures of Galileo, the two witnesses, and Segizzi were not needed, since the signature of the notary, on the first page of the file, guaranteed the juridical validity of all the documents contained in it. On the other hand, the very discrepancies between the instructions given by Cardinal Millini and the document that contains the Segizzi injunction appear to be a in favor of the latter's authenticity. In fact, if one had wanted to create a false document, it should have been fabricated in accord, not in contradiction, with those instructions. That is, one should have said that Galileo had refused to submit himself to the admonition of Bellarmine and that, as a consequence, the commissary had to intervene with a formal injunction.

But what was then the reason for Segizzi's intervention? It is possible to surmise that Galileo, after having heard the admonition of Bellarmine, would have hesitated a moment to respond or made some objection. This is all too natural, since he had gone to Bellarmine's residence unaware of the reasons for his summoning. It was then that Segizzi, already upset by the too moderate manner in which Bellarmine had issued the admonition (as is seen to have been the case from Galileo's deposition at his trial in 1633), could no longer restrain himself and, taking advantage of that moment of uncertainty, would have given the injunction in its more severe form. Faced with such an injunction, Galileo must have promptly submitted. The premature and unjustified intervention of Segizzi, however, must have surely displeased Bellarmine, who considered the phases of the procedure entrusted to him not to have been completed. This is why in the meeting of March 3 he limited himself to affirm that Galileo had been "warned" and that he had also "acquiesced." As far as the cardinal was concerned, Segizzi's intervention had been against the instructions and therefore he omitted mentioning it. On the other hand, Segizzi, who most probably was also present—as customary—at the meeting, well aware of the fact that his intervention had been unjustified, did not dare to give details on what had really happened. And this can satisfactorily explain the difference in the content of the two documents.

As already mentioned, in the following part of the document on the meeting of March 3 the decision of the Congregation of the Index was reported as follows:

> and the decree of the Congregation of the Index having been presented, in which were prohibited and suspended, respectively, the writings of Nicolaus Copernicus *On the Revolutions of the Heavenly Spheres,* of Diego de Zuñiga *On Job,* and of the Carmelite Father Paolo Antonio Foscarini, His Holiness [the pope] ordered that the edict of this suspension and prohibition, respectively, be published by the Master of the Sacred Palace. (Galileo, *Opere,* 19:278; trans. Finocchiaro, *Galileo Affair,* 148)

In fact, the decree of the Congregation of the Index was published two days later (March 5, 1616). After reporting the prohibition of several other works, the decree added:

> This Holy Congregation has also learned about the spreading and acceptance by many of the false Pythagorean doctrine, altogether contrary to the Holy Scripture, that the earth moves and the sun is motionless, which is also taught by Nicolaus Copernicus's *On the Revolutions of the Heavenly Spheres* and by Diego de Zuñiga's *On Job.* This may be seen from a certain letter published by a certain Carmelite Father, whose title is *Letter of the Reverend Father Paolo Antonio Foscarini, on the Pythagorean and Copernican Opinion of the Earth's Motion and Sun's Rest and on the New Pythagorean World System* (Naples: Lazzaro Scorig-

gio, 1615), in which the said Father tries to show that the above-mentioned doctrine of the sun's rest at the center of the world and the earth's motion is consonant with the truth and does not contradict Holy Scripture. Therefore, in order that this opinion may not creep any further to the prejudice of Catholic truth, the Congregation has decided that the books by Nicolaus Copernicus (*On the Revolutions of Spheres*) and Diego de Zuñiga (*On Job*) be suspended until corrected; but that the book of the Carmelite Father Paolo Antonio Foscarini be completely prohibited and condemned; and that all other books which teach the same be likewise prohibited, according to whether with the present decree it prohibits, condemns, and suspends them respectively. In witness thereof, this decree has been signed by the hand and stamped with the seal of the Most Illustrious and Reverend Lord Cardinal of St. Cecilia, Bishop of Albano, on 5 March 1616. (Galileo, *Opere*, 19:323; trans. Finocchiaro, *Galileo Affair*, 149)

Some important documents, recently found in the archives of the Holy Office, throw light on what was happening behind the scenes of this decree. From them we now know that it was once again Bellarmine who received the task of placing before the cardinals of the Congregation of the Index the problem as to what measures should be taken with respect to the Copernican works. The meeting took place on March 1, in the residence of the Jesuit cardinal. The discussion must have been a protracted and heated one, to judge from the minutes: "and after a mature discussion among the above nominated Very Illustrious Cardinals on this matter, finally they decided." We also know from another document that it was the pope who wished that the suspension and prohibition of the Copernican works would not be published separately but together with those of other works, as in fact occurred. Paul V probably wanted to avoid in this way that too much sensationalism be created concerning the decision that had been taken.

It is important to notice that in this decree of the Index the words "heresy" and "error in faith," used by the "qualifiers" of the Holy Office, are absent. The reason for the condemnation of the "Pythagorean doctrine" is simply that such doctrine is "false and altogether opposed to Holy Scripture." The explanation of this absence is given by a friend of

Galileo, Gianfrancesco Buonamici, who was in Rome at the time of Galileo's trial, with the following words:

> In the time of Paul V this opinion was opposed as erroneous and contrary to many passages of Sacred Scripture; therefore, Paul V was of the opinion to declare it contrary to the faith; but through the opposition of the Lord Cardinals Bonifacio Caetani and Maffeo Barberini, today Urban VIII, the pope was stopped right at the beginning on account of the good reasons taken by their Eminences and the learned writing that Mr. Galileo on this matter addressed to the Lady Christina of Tuscany about the year 1614. (Galileo, *Opere,* 15:11)

The last statement of Buonamici is, however, altogether unlikely. As we know, the *Letter to the Grand Duchess Christina* was not made public by Galileo for reasons of prudence. As to the other statements of Buonamici, some confirmation of them may possibly be obtained from what Urban VIII himself declared to Tommaso Campanella in 1630: "It was never our intention [to prohibit Copernicanism]; and if it had been left to us, that decree [of the Congregation of the Index] would not have been made" (Galileo, *Opere,* 14:88). What seems to be certain, however, is that Barberini and Caetani were able to avoid the complete prohibition of Copernicus's *On the Revolutions,* as can be found from some autobiographical notes dictated by Urban VIII. In these notes, the pope adds that Bellarmine, "after having consulted the Jesuit geometers," greatly approved the idea." Caetani's intervention seems furthermore confirmed by the fact that he was put in charge of the corrections to be made in *On the Revolutions.* It is possible that he had spoken, during the meeting, in favor of the scientific value of Copernicus's book.

In the absence of the qualifications of "heresy" and of "error in faith," the declaration of the decree of the Index on the opposition of the Copernican opinion to scripture remained vague, leaving open the possibility of different interpretations. The "rigorists" could still maintain that such an opposition implied heresy or at least an error in faith. For the "moderates," on the contrary, one could only speak of rashness (*temerarietas*). And such ambiguity would continue until the time of Galileo's trial. As we shall see, Urban VIII, who had until then supported the more moderate interpretation, would on that occasion change his opinion in favor of the interpretation of the Copernican thesis as an error in faith, thus leading to the condemnation of Galileo as being "vehemently suspected of heresy" and consequently to his abjuration.

This decree of the Index brings to an end that which is often called the first trial of Galileo. In fact, even though at the beginning there were denunciations against him and against his writings, the conclusion of the affair left aside the person of Galileo, at least in public.

The ambassador Guicciardini, who right from the beginning was against Galileo's coming to Rome, could now show the grand duke that he had been right. He had, as a matter of fact, already done so on the day before the publication of the decree of the Index, in the letter of March 4 to which I have already referred. In the continuation of it Guicciardini gave the following comment on the events:

> Galileo has relied more on his own counsel than on that of his friends. The Lord Cardinal del Monte and myself, and also several cardinals from the Holy Office, had tried to persuade him to be quiet and not to go irritating the issue. If he wanted to hold this Copernican opinion, he was told, let him hold it quietly and not spend so much effort in trying to have others share it. Everyone fears that his coming here may be very prejudicial and that, instead of justifying himself and succeeding, he may end up with an affront. He is all afire on his opinions and puts great passion in them, and not enough strength and prudence in controlling them; so that the Roman climate is getting very dangerous for him, and especially in this century, for the present Pope, who abhors the liberal arts and this kind of mind, cannot stand these novelties and subtleties; and everyone here tries to adjust his mind and his nature to that of the ruler. Galileo has monks and others who hate him and persecute him, and, as I said, he is not at all in a good position in a place like this, and he might get himself and others into serious trouble. (Galileo, *Opere*, 12:241–42; trans. Santillana, *The Crime of Galileo*, 119)

Obviously Guicciardini was exasperated and wanted to persuade the grand duke to recall Galileo to Florence as soon as possible. But Galileo was not the kind of man who quits easily. The decree of the Congregation of the Index had not mentioned him or any of his writings. And concerning the summoning by Bellarmine (which could not have escaped the attention of Guicciardini) Galileo must have limited himself to tell the latter that the cardinal had informed him about the forthcoming decree of the Index on the Copernican works. But he certainly had taken good care not to tell him of the admonition received and even less so of Segizzi's injunction. Therefore, it was not advisable to withdraw precipitously from Rome and thus to give the impression of a personal defeat. Nor, obviously, did Galileo agree with Guicciardini's version of the facts. In fact, writing to Picchena on March 6, he said:

> As one can see from the very nature of the business, I have no interest whatsoever in it, nor would I have gotten involved in it if, as I said, my enemies had not dragged me into it. What I have done on the matter can always be seen from my writings pertaining to it, which I save in order to be able to shut the mouth of malicious gossipers at any time, and because I can show that my behavior in this affair has been such that a saint would not have handled it either with greater reverence or with greater zeal towards the Holy Church. This perhaps has not been done by my enemies, who have not refrained from any machination, calumny, and diabolic suggestion, as Their Most Serene Highnesses and also Your Lordship will hear at length in due course. (Galileo, *Opere,* 12:244; trans. Finocchiaro, *Galileo Affair,* 150–51; slightly modified)

Galileo was certainly well aware of the ambassador's hostility towards him, and he guessed that Guicciardini must have sent to the grand duke reports altogether unfavorable to him. He was thus trying to ward off the blow beforehand, by defending his behavior and by minimizing the importance of the decree of the Index. In the same letter, he remarked to this end that the Copernican theory had not been condemned as heretical and that only those books had been prohibited that intended *ex professo* to prove that that theory was not repugnant to scripture. And he added:

As for the book of Copernicus himself, ten lines will be removed from the Preface to Paul III, where he mentions that he does not think such a doctrine is repugnant to Scripture; as I understand it, they could remove a word here and there, where two or three times he calls the earth a star. The correction of these two books has been assigned to Lord Cardinal Caetani. There is no mention of other authors. (Galileo, *Opere,* 12:241; trans. Finocchiaro, *Galileo Affair,* 150)

As one can see, Galileo was well informed. In fact, the task of introducing the corrections in the *On the Revolutions* had been given to Cardinal Caetani. After his death (June 1617) the task would be taken up by Ingoli, who for a little more than one year had been a consultor to the Congregation of the Index. Although an adversary of the Copernican vision of the world, Ingoli (in agreement with Caetani) was persuaded of the great usefulness of *On the Revolutions* as a mathematical work. He would present his suggestions concerning the corrections in the meeting of the Congregation of the Index on April 2, 1618 (held once more, at Bellarmine's residence). The cardinals' decision at that meeting would be to put the proposed corrections to the mathematicians of the Roman College. Bellarmine would assume the job, entrusting the examination of the question to Fathers Grienberger and Grassi. They gave a fully positive opinion, and thus Ingoli's plan for corrections was approved on the following July 3. However, it would still take almost two years before the publication of the decree of the Congregation of the Index on May 15, 1620, that allowed the publication of the corrected edition of *On the Revolutions.* As a matter of fact, however, such an edition would never see the light of day, and thus this work remained on the Index of Forbidden Books until the year 1835. Ingoli would carry on a parallel action for the prohibition of Kepler's *Epitome Astronomiae Copernicanae,* which would come to an end one year before (May 10, 1619) with the unconditional prohibition of this work. Kepler would deplore such a prohibition, as due to the "inappropriateness of some who have treated of astronomical matters in places where they should not have been treated and with improper methods," with a probable allusion, in addition to Foscarini, to Galileo himself (Paschini, *Vita e Opere,* 354).

That the Church had wanted to treat Galileo with special regard is shown by the audience granted to Galileo by Paul V just one week after the decree of the Index was issued. According to the report of Galileo to Picchena, the pope stayed with him at length ("three quarters of an hour"), showing himself quite benevolent and assuring him that he was convinced of his intellectual integrity and sincerity. And Galileo continued:

> Finally, since I appeared somewhat insecure because of the thought that I would be always persecuted by their implacable malice, he consoled me by saying that I could live with my mind at peace, for I was so regarded by His Holiness and the whole Congregation that they would not easily listen to the slanderers, and that I could feel safe as long as he lived. Before I left he told me many times that he was very ready at every occasion to show me also with actions his strong inclination to favor me. (Galileo, *Opere*, 12:248; trans. Finocchiaro, *Galileo Affair*, 152)

In spite of Ambassador Guicciardini's pressuring for a recall of Galileo to Florence, Galileo remained in Rome under various pretexts until the beginning of June. Undoubtedly, he wanted also to ascertain what had become known about his summoning by Bellarmine. That summoning, as well as the fact that the commissary of the Holy Office had been present at the meeting, could have not have escaped public attention. After a little more than a week, the decree of the Index had come out. Given the extreme secrecy surrounding the Holy Office proceedings, nothing could have been known with certainty about what had taken place in Bellarmine's residence. It was, however, all too natural that rumors would start to spread. According to them, Galileo had been called by the cardinal on orders of the Inquisition to give an account of his Copernican convictions, which he had been obliged to abjure. After that, severe penances had been imposed on him by Cardinal Bellarmine. Such rumors had very soon even reached Venice, as was evident from the reports sent to Venice on February 27 and March 12 by Contarini, the ambassador of the Republic of Venice in Rome, as well as from a letter sent to Galileo on March 11 by his friend Sagredo.

Since there was no indication that these rumors were going to die out, Galileo decided to have recourse to Bellarmine himself. On May 26 Bellarmine released to him the following declaration:

> We, Robert Cardinal Bellarmine, have heard that Mr. Galileo Galilei is being slandered or alleged to have abjured in our hands and also to have been given salutary penances for this. Having been sought about the truth of the matter, we say that the above-mentioned Galileo has not abjured in our hands, or in the hands of others here in Rome, or anywhere else that we know, any opinion or doctrine of his; nor has he received any penances, salutary or otherwise. On the contrary, he has only been notified of the declaration made by the Holy Father and published by the Sacred Congregation of the Index, whose content is that the doctrine attributed to Copernicus (that the earth moves around the sun and the sun stands at the center of the world without moving east to west) is contrary to Holy Scripture and therefore cannot be defended or held. In witness whereof we have written and signed this with our own hands, on this 26th day of May 1616. (Galileo, *Opere,* 19:348; trans. Finocchiaro, *Galileo Affair,* 153)

It is interesting to note that Bellarmine did not speak of the acquiescence by Galileo. He also did not mention the intervention of Commissary Segizzi, a circumstance that has seemed to several Galileo scholars to be a confirmation of the fact that the document found in 1632 and mentioned here above, is a fraud. However, Bellarmine's declaration does not appear to be in clear contradiction with that document. In fact, Bellarmine simply stated that "he has only been notified," without adding "by us." Such a rather vague formulation does not exclude, therefore, the Segizzi intervention. As I have already surmised, such an intervention must have been considered by Bellarmine as premature, and this is why he does not mention it here, as he had not mentioned it on the occasion of the meeting of the Holy Office of March 3.

Galileo would jealously keep this declaration and would make ample use of it, as we will see, on the occasion of his trial. In the meanwhile, Picchena had sent to Galileo a courteous but clear invitation to return to Florence as soon as possible, and Galileo started his return trip at the beginning of June. Besides the precious attestation of Bellarmine, he brought with him also those of Cardinals del Monte and Orsini. The latter emphasized the fact that Galileo was leaving Rome keeping intact his reputation and the esteem of all of those who had dealt with him. And they gave assurance that it was clear to everybody how his enemies had maliciously spoken against him.

But despite those attestations Galileo must have felt a deep bitterness. The plan he had conceived to silence the opposition of his enemies by convincing the Church authorities that they must not hastily judge the Copernican issue had failed. And no small part of that deep bitterness must have been directed towards the Jesuits. Galileo had hoped for active support from them, if not on the issue of Copernicanism, at least to avoid a premature condemnation of Copernicanism. They, however, had not acted. Sagredo, Galileo's best Venetian friend, had written to him on March 11, 1616, insinuating that the Jesuits had not limited themselves to a nonintervention policy, but had even carried on a positive action in favor of putting *On the Revolutions* on the Index of Forbidden Books. Sagredo had always been very critical of the Jesuits, which Galileo must have known. But undoubtedly, in the state of mind in which he found himself now, word of a friend must have affected him, greatly increasing his resentment. As we will see, such resentment will not fail to become evident within a few years and will progressively bring to a complete breaking off of Galileo's relationship with the Jesuits, which previously had been marked by the highest reciprocal esteem.

Many biographers of Galileo have also severely criticized the Jesuits for not having taken a supporting role in favor of Galileo at that critical moment. Others, on the contrary, tend to highlight the fact that the reservation of the Jesuits is to be attributed (at least for the most part) to "scientific" reasons. The Jesuits were by now convinced, as Clavius himself had admitted at the end of his life, that the Aristotelian-Ptolemaic system could not be upheld. They were also aware, however, that there was a lack of valid proofs in favor of Copernicanism. Thus, they preferred to wait, adopting for the time being Tycho Brahe's system, which avoided problems with scripture.

In order to evaluate correctly the Jesuits' attitude, it seems to me necessary to reconsider their power to influence events. Certainly, thanks to the special characteristics impressed upon it by its founder, Ignatius of Loyola, the religious Order of the Jesuits had right from the beginning an important role in the Catholic Counter-Reformation, and for that reason it even enjoyed the favor of many popes. This role continued to increase in depth and in breadth, thanks especially to the Jesuits' activity in the field of education. Already at the beginning of the seventeenth century the schools (or "colleges") of the Jesuits were forming the social and intellectual elite of Catholic Europe.

But alongside the Jesuits there also existed other older religious orders, the first among them being the Dominicans. They boasted of a very rich theological tradition, from St. Thomas of Aquinas onwards. And the fact that in theology, the Dominicans, indeed, were not disposed to concede to the Jesuits the defense of Catholic orthodoxy is shown by the long and bitter theological controversy *de auxiliis,* concerning the reconciliation of divine grace with human free will. The controversy began in 1599 and went on until 1607, with the most renowned Dominican and Jesuit theologians disagreeing with one another. And it came to an end without a victor or vanquished with the order by Paul V (at the suggestion of Bellarmine) to put an end to the discussions in expectation of a decision by the pope on the matter, a decision which in fact was never made.

Having learned from this experience, the Jesuits were undoubtedly aware of the limits of their influence, even if they had wanted to carry out activities in Galileo's favor. Such an initiative, in any case, should have been taken by the mathematicians of the Roman College. Now, the latter made up only a small minority with respect to their fellow religious teachers of philosophy and theology. The philosophy teachers were certainly not narrow-minded Aristotelians. And the insistence of the superior general of the Order, Acquaviva, in his letters of 1611 and 1613 addressed to the Jesuit professors of philosophy and theology on the necessity to keep to the Aristotelian doctrine, shows that they were not doing so enough. But they did not have available for their teaching of philosophy anything other than the great Aristotelian synthesis "christianized" during the Middle Ages. There did not yet exist a new natural philosophy, much less a comprehensive philosophical synthesis. The man who is often called the father of modern philosophy, René Descartes (1596–1650), had just two years earlier, in 1614, completed his studies in the famous college of the Jesuits, La Flèche. As to Francis Bacon (1561–1626), another figure central to the era's change in scientific practice, he was just then putting together the various drafts of his *Novum Organum,* finished much later, in which were laid the foundations for the new empirical method to replace that of Aristotle. In short, modern philosophy was still in its gestation period. And Galileo himself, who was the one moving most consciously in the direction of a new natural philosophy, was far from possessing (and from being able to offer) a unified view of what would one day come to birth.

Given such a situation of uncertainty and disorientation, it was inevitable that at the Roman College, as in all European universities of the period, the Protestant ones included, Aristotelian philosophy continued to be taught. Moreover, in the case of the Jesuits, there were the prescriptions of the constitutions and of the other official documents of the Order, as well as the letters of General Acquaviva.

As to the theology professors, the reconciliation of Copernicanism with scripture weighed more heavily upon them than did the problem of Aristotelianism. Even though they were at times more open than their colleagues in philosophy, the Jesuit theologians were certainly not capable of following in the wake of Father Foscarini. And the extreme reserve (to say the least) that Bellarmine had shown with respect to future proofs of Copernicanism had certainly an effect on their attitude too.

If one takes all this into consideration, it is much easier to understand the attitude of Grienberger and the other mathematicians of the Roman College. By now there were among them some who were in their own hearts Copernican. Others, like Grienberger himself, were at least strongly inclined towards heliocentrism. But there remained always some doubts. And the prescriptions of their father general acted as a further restraint to any pro-Galilean activity by them. Under those conditions, how could they have been able to advance the ideas of Galileo? There was nothing left but to hope (and even perhaps to pray) that Galileo or others would finally succeed in providing convincing and irrefutable proofs of Copernicanism. In this case, no one, not even their father general or Paul V, would have been able any longer to force them to a "blind obedience" in favor of Aristotelianism.

Even the responsibility of Bellarmine in the decisions of 1616 is in need, I think, of a more objective appraisal. Surely Bellarmine was one of the most authoritative Catholic theologians of the epoch, and his influence was undoubtedly profound. And from the documentary information that has been examined up to the present, it seems that we can draw the conclusion that Bellarmine was certainly in agreement with the publication of the decree of the Index that sanctioned the opposition of Copernicanism (as a physical worldview) to Holy Scripture. How could one reconcile this fact with Bellarmine's admission, in his response to Foscarini, of a possibility, even if only a theoretical one, that one day Copernicanism could be proved? In fact, that decree of the Index excluded such a possibility peremptorily and definitively. Bellarmine could have suggested a suspension of judgment, so that the decree would have been avoided, as it seems Cardinals Barberini and Caetani tried to do. Why did Bellarmine not do the same?

Even admitting all this, however, it was certainly not Bellarmine himself who imposed, on the basis of a personal decision, the line to be taken in the whole question. The preoccupation with the defense of Catholic orthodoxy against all the various "heresies" of the epoch was too widespread at the highest levels of the ecclesiastical hierarchy to allow for a different solution, even without Bellarmine's influence. And in the Holy Office Galileo had powerful opponents, one of whom was probably the Dominican Cardinal Gallamini, who appears to have been the supporter, if not the promoter, of the actions of Caccini against Galileo.

As to Galileo's responsibility for the prohibition of Copernicanism, it has very often been emphasized, starting from Ambassador Guicciardini up to our own day, especially in apologetic writings by Catholics. Galileo, one affirms, lacked prudence and tried to trick the Catholic Church into accepting Copernicanism even though he did not have convincing proofs for it. Moreover, he wanted to give a theological answer to the thorny question of how to reconcile the Copernican view with scripture, despite the fact that he was a simple layman. As a result of this untimely zeal and of this imprudent intrusion into the biblical field, he precipitated a decision of the Church that could have been otherwise avoided.

In reality, Galileo showed himself on several occasions to be more prudent than his admiring friends. To be sure, his trip to Rome in 1615–16 ended by revealing itself to have been a serious tactical error. But it is still true that such a trip was motivated by the ever-growing concern that his enemies would succeed in provoking a decision against Copernicanism within a short time. Given their temperament and the means they used, such a concern was certainly not unfounded. As to the accusation that he had the intention to extract from the Roman authorities with all available means a decision favorable to Copernicanism does not seem to correspond to the facts. As can be clearly established from the *Letter to Castelli* and from the two letters to Dini, and above all from the *Letter to the Grand Duchess Christina,* Galileo was completely aware of the prudence required in order that the Church might come to a conclusion about Copernicanism. In fact, what he sought to obtain in 1615–16 was that the Roman Church authorities would leave the Copernican question open to discussion, without making a hasty decision in the matter.

Even Galileo's intrusion into the scriptural field was motivated by the same concern about such a hasty decision by the Church. As we have seen, it was in fact his adversaries who, more and more at a loss to battle Galileo in the field of "natural philosophy," had been the first ones to use biblical passages in an anti-Copernican campaign. What should Galileo have done? Should he have remained quiet and strengthened his case with scientific arguments, as suggested by Grienberger, among others? To Galileo, however, to remain silent would have meant that the scriptural argument was by itself alone sufficient to solve a question that instead, according to him, should have been left open to discussion.

On the other hand, the question still remains as to whether, had Galileo not intervened in the field of scripture, the Church would have avoided taking an official position. Foscarini had intervened, totally independently from Galileo. As we know, Ciampoli, quite a bit more perspicacious than Cesi, had foreseen that the *Letter of Foscarini* would be condemned long before Galileo's trip to Rome. And that condemnation could not have come about without having been motivated by the opposition between Copernicanism and Holy Scripture, namely, without a stance having been taken such as the one that came about with the decree of the Index in 1616. Thus, quite apart from Galileo's activity, the question of Copernicanism's correspondence to scripture was likely to have arisen.

And as for the fact, repeated very often by Catholic apologists in the past and even today, that Galileo had not brought forth decisive proofs for Copernicanism, it seems to me that it is necessary to be more specific about such "lack of proofs." Surely enough, Galileo was never able to give fully convincing proofs before the decisions of 1616, nor even afterwards. In order to reach a theoretical justification of the Copernican system of the world, the development of the Newtonian physics would be required. To its foundation, Galileo, together with Kepler, had contributed, but it will be brought to completion only seventy years later, by Newton. As to observational support for Copernicanism, in Galileo's time telescopes were not adequate to observe stellar parallax, which would be detected for the first time more than two centuries later. However, from the totality of the undisputed data obtained from his observations, as well as from those of other astronomers, including the Jesuits themselves, Aristotelian cosmology and Ptolemaic astronomy had been irremediably put in crisis. There was, it is true (even though, at the time, within a very restricted field of specialists) Tycho Brahe's geoheliocentrism. At any rate, the minimum that could be said was that Copernicanism appeared by now as a real possibility. That was sufficient authorization for Galileo to recommend to the Church that it not jump precipitously into a negative decision with respect to that theory.

None of this, unfortunately, was taken into consideration by the qualifiers of the Holy Office. It seems to me important to stress, once more, that they did not even suggest to themselves the problem of whether or not scientific proofs existed to support Copernicanism, let alone their value. They were, in fact, convinced that they possessed already the philosophical truth, that is, that of the Aristotelian-Thomistic natural philosophy. Such a conviction made them certain a priori that there did not exist then, nor could there ever exist in the future, any scientific argument contrary to that philosophy. Each one of the two Copernican propositions submitted to their judgment was therefore, as they declared, "foolish and absurd in philosophy." It was indeed such philosophical certainty on their part that authorized them to conclude, theologically, that those propositions, especially the first one, on the Sun's immobility, were in contradiction with Holy Scripture.

One can, therefore, conclude that at the basis of the certainty with which in 1616 the Church rejected Copernicanism (with an intention that it be definitive) was not only the theology of the epoch, but also, and first of all, the philosophy that was so closely linked to the theology as to constitute an inseparable whole. Surely, the enormous difficulty, both intellectual and psychological, of making theology independent of a worldview based on the obdurate convictions of common sense, and which for centuries had been considered as intimately linked to the theological Christian synthesis, may provide extenuating circumstances for the erroneous decision of 1616. Another attenuating circumstance may be seen in the concern that the new ideas being spread by Galileo, with which even a theologian like Foscarini had associated himself, might present a serious danger for the unity of the Catholic faith, together with the challenge to the teaching authority of the Church. But even granted these circumstances, there is still the grave objective error of having wished to resolve authoritatively and definitively a question that should have been left open, and to have sought to silence in the same definitive manner those who promoted the new ideas. As we shall see, this abuse of power both doctrinal and disciplinary will have its inevitable sequel in the trial and condemnation of Galileo in 1633. And it will continue for centuries to weigh heavily upon the relationship of the Church with modern thought.

From the Polemics on the Comets to the *Dialogue*

Upon his return to Florence, Galileo did not allow himself to be discouraged. He was certainly well aware of the necessity to keep an attitude of prudent silence on Copernicanism, bounded as he was by the admonition received from Bellarmine and that much more severe from Segizzi. At the same time, optimist by temperament as he was, he must have undoubtedly hoped that, with the lapse of time and with the possibility of finding convincing proofs in favor of the Copernican system, the Church's attitude could be changed.

Making use of the peace and the free time that was once more his, after the years of frantic polemical activity, he returned to his long-standing studies on motion by reorganizing and selecting the materials from his time at Padua. And he dedicated himself also to astronomical observations. But it was a slow undertaking, broken up by periods of sickness, which pushed him to move to the villa of Bellosguardo on the hills surrounding Florence.

Despite his good intentions, Galileo's silence was not to last long. Towards the end of the year 1618 three comets appeared over a short period of time. The third and largest one, in particular, did not fail to arouse the deep impression that always accompanied the appearance of those heavenly phenomena, interpreted as a forewarning of cataclysms and of wars. In fact, the appearance of these comets coincided with the beginning of the long war that subsequently became known in history as the Thirty Years War, a fact that seemed to confirm such popular beliefs.

As usual, extremely heated discussions arose among astronomers and natural philosophers. According to the Aristotelian explanation, adopted by most of the latter, the comets were terrestrial exhalations that, as they rose to the highest zone of the sphere of fire, became heated by the friction caused by the movement of the sphere of the Moon, immediately above it. And they were made to rotate in a circle by effect of the same circular motion of the latter. Their disappearance coincided with the total extinction of their combustible material. Others, and especially the astronomers, subscribed instead more or less faithfully to the opinion of Tycho Brahe. According to it, the comets were located quite a distance beyond the Moon and moved around the Sun in an orbit which was probably not circular but oval, close to the orbit of Venus. And that explained their varying distance from the Earth. According to Brahe, comets were a kind of transient celestial phenomenon, as were the novae, localized, not in the sphere of the "fixed stars" as the latter were, but in that of the planets.

Galileo was prevented from observing these comets because of a sickness that forced him to stay in bed during the entire period that they remained visible (see Galileo, *Opere,* 6:225). On the other hand, he was persuaded (and remained so until the very end of his life) that the phenomenon of the comets was difficult to understand, and thus he preferred to remain silent.

A public conference on the comets was held at the Roman College by Father Orazio Grassi (1583–1654), who, at that time, held the chair of mathematics in the place of Grienberger. That conference was published in the following year with the title *An Astronomical Discussion of the Three Comets of 1618,* but without Grassi's name, for reasons of prudence. Grassi, in fact, defended an explanation of the phenomenon that could have stirred up polemics with the Aristotelians: he affirmed that the comets were far beyond the Moon, and presumably between the latter and the Sun. In that way, he sided with Brahe and against Aristotle. Concerning the center of the comet's orbit, however, he seemed to consider it to coincide with the Earth and not the Sun, in accordance, in this instance, with Aristotle.

Galileo received in advance the news of the printing of the *Astronomical Discussion,* at the beginning of March 1619, from one of his Roman correspondents, Rinuccini, who wrote to him:

> The Jesuits presented publicly a Problem [on the distance of the comet] which has been printed and they hold firmly that it is in the sky [that is, beyond the Moon], and some others than the Jesuits have spread it around that this thing overthrows the Copernican system, against which there is no surer argument than this. (Galileo, *Opere,* 12:443; trans. Drake, *Galileo at Work,* 265)

Rinuccini insisted that Galileo enter into the debate and present his opinion. Rinuccini's hint of the phenomenon of the comets having been used against Copernicanism had been vague. Recent research has furnished evidence of the fact that lively discussions on the comets took place within a group of Roman intellectuals who gravitated around Cardinal Scipione Corbelluzzi. He had put to them the question whether from the motion of the comet one could in any way draw an argument against the Copernican motion of the Earth. And Francesco Ingoli, who was a member of the group, had answered that question, in a fully affirmative way, with an unpublished writing, *Treatise on the Comet of 1618.* As justification, he used the same argument already used by Brahe against Copernicanism, namely, that of the (asserted) absence of variation of cometary parallax during the comet's motion around the Sun. On the contrary, such variation should have become visible if one assumed a motion of the Earth around the Sun.

We do not know whether Galileo had any more precise information on these discussions in Roman intellectual circles. But just the hint from Rinuccini was enough to give him serious concerns. On the other hand, once he had read the *Astronomical Discussion* of Grassi, he undoubtedly noticed that the latter had based his arguments mostly on Tycho Brahe's theory of comets. Now, Galileo must have certainly known

that Brahe had put forth his theory precisely because of the difficulties, for him insurmountable, which comets presented for Copernicanism. Even if not explicitly anti-Copernican, as those of Ingoli, Grassi's affirmations were such, as a matter of fact.

Obviously, Galileo was not allowed to take an open position in favor of Copernicus. But he could do so indirectly, by showing the shortcomings of the cometary theory of Brahe, as well as of that of his "disciple" Grassi, thus neutralizing its use for an anti-Copernican purpose. So he decided to intervene. As a precaution, he preferred to have Mario Guiducci speak in his place. Born in Florence in 1585, Guiducci had studied at the Roman College. Before 1618 he had worked as Galileo's assistant, transcribing the latter's notes from his Paduan period. The conference on the comets, held by him at the Florentine Academy, was published towards the end of June, with the title *Discourse on the Comets of Guiducci*. In fact, as is apparent from the manuscript, this *Discourse* was composed for the most part by Galileo himself.

Guiducci (understood largely as a mouthpiece for Galileo) criticized first of all the Aristotelian theory of the comets, as one full of absurdities. But he was no less severe towards that of Tycho Brahe, which had been taken up by the author of the *Astronomical Discussion* (Guiducci never mentioned the name of Grassi). According to Guiducci, the argument for parallax could not be applied to the comets, unless one had first proven that they are true material bodies. In like manner he criticized the opinion of Tycho Brahe of a motion (perhaps not even circular) of the comets around the Sun or, as in the case of the mathematicians of the Roman College, of their circular motion around the Earth. After this critical review of the preceding opinions, Guiducci put forth Galileo's opinion in a manner of a more likely hypothesis but without pretending to give a solution to the problem.

According to this opinion, the comets could be optical phenomena caused by the reflection of sunlight by the exhalations of vapors rising vertically from the Earth and extending to the regions beyond the Moon. Even though Galileo seemed thus to be recalling the opinion of Aristotle, as to the origin of the material of the comets, he detached himself radically from that opinion for three reasons: (1) the light of the comets did not come from a "fire" of such exhalations; (2) their motion was not circular; and (3) above all else, the exhalations of the Earth were capable of rising from the terrestrial region of the four elements, right up to the celestial region above the Moon. Such an affirmation was in clear contradiction to the radical distinction between the earthly and the celestial bodies according to Aristotle.

Having once risen to that height, these exhalations were illuminated by sunlight and in that way became visible and thus later on disappeared, due to the progressive movement away (always in a straight line). That it was a matter of "vapors" and not of real true heavenly bodies seemed confirmed, according to Guiducci, by the fact that it was possible to observe stars through the comets.

A serious contradiction remained in the whole argumentation of Guiducci on parallax. On the one hand, he was putting in doubt that the latter could be used in the case of comets, in order to show that they were true celestial bodies. On the other hand, he affirmed that the illumination of the terrestrial exhalations happened beyond the Moon, as proved by the "smallness of the parallax observed with utmost care by so many excellent astronomers." It is truly strange that Galileo was not aware of this contradiction, which Grassi naturally did not fail to make evident in his reply (as described below).

Guiducci admitted that there remained only one problem in his explanation. Since the exhalations were going up with a motion perpendicular to the surface of the Earth, they should have appeared directed towards the zenith. In fact, however, their motion appeared "inclined towards the North." This fact, continued Guiducci, compels us either to abandon the hypothesis, "even though it corresponds to the appearances in so many cases or to add some other cause for this apparent deviation." And he added:

> I would not be able to do the one, nor should I venture to do the other. Seneca was aware and he wrote how important it was to have a sure determination of these things, to have a solid and unshakable knowledge of the order, of the disposition, and of the states and movements of the parts of the universe, a knowledge which is lacking to our century; however, we should be content with that little bit that we can conjecture amidst the shadows, until we are told the true constitution of the parts of the world, because what Tycho has promised remains imperfect. (Galileo, *Opere,* 6:98–99)

With these words, Guiducci made clear the difficulty of explaining the phenomenon of the comets in an age like his, when one remained without a credible cosmology which would supply the means to do it. Aristotelian cosmology was evidently false. That of Tycho Brahe remained "imperfect." Copernican cosmology was left, but of that (which for Guiducci was certainly the only true one) it was not allowed to speak. The only course left, therefore, was to try to thus "conjecture amidst the shadows."

In substance, Galileo had wished, through Guiducci, rather than to give a theory of the comets, to show the insufficiency of the explanation of Tycho Brahe, and, therefore, of all those who followed him, as did the Jesuit mathematicians of the Roman College. This was enough to show the vanity of the pretense that such an explanation was able to demolish the Copernican theory. It was this, above all else, that had pressed Galileo to intervene.

But there were also, in that intervention, expressions of personal resentment. As I have already said, Grassi had deduced in the *Astronomical Discussion* that comets were a superlunary phenomenon on the basis of the smallness of their parallax with respect to that of the Moon. But he had also indicated, as a confirmation of his thesis, a lesser magnification of them, by the telescope, than that of the Moon. According to Grassi, in fact, the magnification of an object by a telescope is the less, the greater is its distance. And Grassi had added: "I know that scarce importance has been attributed to this argument by those who pay little attention to the principles of perspective." Galileo (who rightly considered that argument as totally false) had been deeply irritated by this phrase, which he interpreted as an accusation of ignorance of the principles of optics addressed against him. And his resentment was clear through his characterization of Grassi's optical theory as "most vain and false."

Moreover, there was no lack here and there in the *Discourse* of critical remarks concerning the scientific posture of the mathematicians of the Roman College. In addition to previous resentments they indicated an increased resentment, nourished in those years of silence, for the policy of nonintervention that the Jesuit mathematicians had followed on the occasion of the events of 1616.

All this could not fail to deeply offend the Jesuits, and Grassi in particular, the more so since everybody had understood that the real author of the *Discourse* was Galileo. There had been, moreover, some stinging remarks concerning Scheiner. Once Scheiner came to know them, he promised to "repay Galileo with his own money," and he would do so abundantly, as we will see.

Ciampoli himself, after having come to know about the *Discourse,* did not hesitate to manifest frankly to Galileo all his concern for the foreseeable estrangement of the Jesuits of the Roman College. There was no waiting for Grassi's response to Galileo. In fact, it appeared in the autumn of the same year under the title *The Astronomical and Philosophical Scale.* The author's name was "Lothario Sarsi Sigesano," which was an anagram of "Horatio Grassi Saloniensi." "Sarsi" presented himself as a disciple of Grassi who wanted to revenge the reputation of his master, which had been injured by Galileo's attack.

In his answer Grassi showed that he was well informed even with respect to the most recent publications, such as those of Kepler. In particular, his comments in the field of optics were often pertinent. Nor was he less justified in pointing out the contradiction into which Galileo had fallen concerning the parallax argument. Furthermore, Grassi offered personal remarks of an experimental character that were certainly of value. At the same time, however, he showed strange trust in the opinions of ancient authors. An example was that of the eggs that the Babylonians are said to have caused to cook by rotating them rapidly in slings. He also presented the opinion of more recent authors who claimed that cannonballs melted during their trajectory through the air because of the heat caused by friction. Furthermore, he at times presented Galileo's statements in a nonobjective way, if not outright distorting them, even though he pretended that he wished to evaluate them impartially.

And there was no lack of biting jokes, at times even insidious. Above all was the one Sarsi made as to the words already quoted of Guiducci on "some other reason" for the apparent deviation towards the north of the straight-line motion of the comet, an explanation, "Guiducci had added," which "I would not venture to do." Grassi had substituted, in his quotation of Guiducci, the words "some other reason" with "some other motion." And he had remarked: "What is this sudden fear in an open and not timid spirit which prevents him from uttering the word that he has in mind?" Wondering thereafter what kind of movement that one could possibly be, he had added:

> I fancy I hear a small voice whispering discreetly in my ear the motion of the Earth. Get thee behind me thou evil word, offensive to truth and to pious ears! You have spoken it with bated breath. For, if it were really thus, one would affirm the opinion of Galileo. If however the Earth does not move, that straight-line motion is not in accord with the observations; for Catholics, however, it is certain that the Earth does not move. But certainly Galileo had no such idea, for I have never known him to be other than pious and religious. (Galileo, *Opere,* 6:145–46; trans. Santillana, *The Crime of Galileo,* 154–55, slightly modified)

After reading this answer, Galileo sought the advice of his Roman friends as to what he should do. They were of the opinion that it would be better if Galileo did not respond directly, so as not to further exacerbate the situation with the Jesuits. And thus it was decided that Galileo would direct his response to the nephew of Cesi, the very young Duke Virginio Cesarini. It was however only in the summer 1621 that Galileo started the preparation of the projected work. He was well aware of the need to be very cautious. It is true that Paul V had died in January 1621, followed by Bellarmine in September of the same year. But on February 28 his protector and faithful admirer, the Grand Duke Cosimo II, had unexpectedly passed away, leaving a son, Ferdinando, who was only eleven years old at that time. The regency of the grand duchy was taken up by the Grand Duchess Christina and by Ferdinando's own mother, Maria Magdalena of Austria. These two ladies of a rigid piety could be influenced through religious arguments by the adversaries of Galileo. And one of the most ruthless among them, Caccini, had returned just a short while before to Florence, and he certainly did not cease to "keep a check" on Galileo.

Galileo's manuscript, entitled *The Assayer,* was finally completed in October 1622 and sent to Cesarini. The latter had various copies of the manuscript made and distributed to members of the Lincean Academy and to Ciampoli, in order to have their comments on it. He afterwards transmitted the manuscript to Cesi, who wrote to Galileo to say that he was reading it "with utmost satisfaction."

Galileo's friends were all of the opinion that the work should be printed in Rome. Cesarini informed Galileo of this decision at the beginning of January 1623, adding that the Jesuits of the Roman College had got wind of the arrival of *The Assayer* in Rome and had asked him for a copy of it. "But," he added, "I have refused them it because they would have been able more effectively to obstruct its publication" (Galileo, *Opere,* 13:106). In the same letter Cesarini informed Galileo that at Rome the sale of the *Apology for Galileo* by Tommaso Campanella, printed the year before at Frankfurt, had been prohibited. Campanella had written this work, as he himself affirmed, at the request of Cardinal Caetani at the beginning of 1616. It seems that the cardinal had wanted to document himself on the vigil of the Church decisions on Copernicanism. But in fact, Campanella's manuscript did not reach him until after the appearance of the decree of the Index.

Cesarini added that in the inaugural lectures of the new academic year, held at the Roman College, the Jesuit professors had spoken against the "finders of novelties in the sciences" and that "with long orations they sought to convince the students that there was no truth outside of Aristotle" (13:107). The review of *The Assayer,* for permission to publish, was entrusted to the Dominican Niccolò Riccardi. On February 1623 he declared that not only had he found nothing erroneous in Galileo's work, but that, on the contrary, he had:

> noted many fine considerations which have to do with natural philosophy thanks to the subtle and solid speculation of the author in whose days I consider myself happy to have been born, when, no longer with the steelyard and roughly, but with such delicate assayers the gold of truth is weighted. (Galileo, *Opere,* 6:200)

With these words, Riccardi had made an allusion to the meaning of "assayer," that is, of a precision scale, that had been used by Galileo in order to compare his accurate measure of the weight of his own arguments and of those of Sarsi, with the rough measure the latter had attempted to achieve with an ordinary steelyard.

While the printing of *The Assayer* was in progress, the successor of Paul V, Gregory XV, took ill and died on July 8, 1623, after only two years of his pontificate. One month later the new pope was elected. It was Cardinal Maffeo Barberini, and he took the name of Urban VIII. That election was greeted with enthusiasm by progressive Catholics throughout Europe. A man of noteworthy intelligence and culture and an able diplomat, he seemed, at age 53, the ideal person to lead the Church in such difficult times as those through which Europe was passing. The difficulties were most serious on the political plane, with the Thirty Years War, which had begun five years earlier, but also profound on the cultural plane. As we have already seen, Urban VIII had shown as a cardinal a notable moderation with respect to the Copernican problem. And he had kept his full esteem for Galileo even after the events of 1616. On August 20, 1620, he had sent him a Latin ode, *Adulatio perniciosa,* some verses of which celebrated Galileo's astronomical discoveries. In the letter that accompanied the ode, Maffeo Barberini requested Galileo to accept it as a "small sign of the great good will that I have towards you" (Galileo, *Opere,* 13:48). On his part, just two months before the cardinal became pope, Galileo had sent him a congratulatory letter on the occasion of the doctorate obtained by his nephew Francesco Barberini. And the cardinal had thanked him on June 24, adding at the end of his letter:

> I remain much obliged to Your Lordship for your continued affection towards me and mine, and I wish to have the opportunity to do likewise to you, assuring you that you will find in me a very ready disposition to serve you out of respect for what you so merit and for the gratitude I owe you. (Galileo, *Opere,* 13:119)

The election of Maffeo Barberini as the head of the Church could thus not but arouse the greatest hopes in Galileo too. And the news that continued to come to him from Rome was such as to fully confirm them. Virginio Cesarini had been appointed as Master of the Chambers of the new pope, and Ciampoli had been confirmed as Secretary of the Briefs to the Princes, a role that in a certain way resembled that of secretary of state in today's Church.

Given the extremely favorable moment, it was decided to dedicate *The Assayer* to the new pope, in the name of the Lincean Academy.

On October 27 the book finally was issued. As Francesco Stelluti announced to Galileo, on that same day a copy of the book had been presented to Urban VIII, and others had been distributed among cardinals and Roman friends. This work of Galileo has been rightly defined as a stupendous masterpiece of polemical literature. And, in fact, the problem of the comets, rather than being the fundamental theme of *The Assayer,* is only its point of departure.

As a result of the polemical character of the book, Galileo had been obliged to follow, point by point, the order of the questions in the work of Sarsi. As a result, he does not propose to us a systematic treatise of scientific methodology or of philosophy of science. Nevertheless, the new methodological and philosophical posture at the basis of his argumentation comes out clearly in its essential features.

Concerning the display of quotations from famous authors, which Sarsi lined up as a support for his own theories, Galileo wrote:

> In Sarsi I seem to discern the firm belief that in philosophizing, one must support oneself upon the opinion of some celebrated author, as if our minds ought to remain completely sterile and barren unless wedded to the reasoning of some other person. Possibly he thinks that philosophy is a book of fiction by some writers, like the *Iliad* or *Orlando Furioso,* productions in which the least important thing is whether what is written there is true. Well, Sarsi, that is not how matters stand. Philosophy is written in this grand book, the universe, which stands continuously open to our gaze. But the book cannot be understood unless one first learns to comprehend the language and the letters in which it is composed. It is written in the language of mathematics, and its characters are triangles, circles, and other geometric figures without which it is humanly impossible to understand a single word of it; without these, one wanders about in a dark labyrinth. (Galileo, *Opere,* 6:232; trans. Drake, *Discoveries and Opinions,* 237–38)

What was really the new natural philosophy advocated by Galileo came out clearly from the considerations developed towards the end of *The Assayer* on the nature of heat. According to Aristotle, motion is the cause of heat. Galileo, through Guiducci, had denied this, thus rejecting the Aristotelian explanation for the origin of comets. Even without accepting the latter, Sarsi had wanted to prove in the *Scale* that, in itself, the Aristotelian principle was correct.

Before explaining in what sense this Aristotelian principle might be true, Galileo wants first to make clear the concept of "hot," and so he comments on it:

> I suspect that people in general have a concept of this which is very remote from the truth. For they believe that heat is a real phenomenon, or property, or quality, which actually resides in the material by which they feel themselves warmed. Now, I say that whenever I consider any material or corporeal substance, I immediately feel the need to think of it as bounded, and as having this and this shape; as being large or small in relation to other things, and in some specific place at any given time; as being in motion or at rest; as touching or not touching some other body; and as being one in number, or few, or many. From these conditions I cannot separate such a substance by any stretch of my imagination. But that it must be white or red, bitter or sweet, noisy or silent, and of sweet or foul odor, my mind does not feel compelled to bring in as necessary accompaniments. Without the senses as our guides, reason or unaided imagination would probably never arrive at qualities like these. Hence I think that tastes, odors, colors, and so on are no more than mere names, so far as the object in which we place them is concerned, and that they reside only in the consciousness [sensitive body]. Hence if the living creature were removed, all these qualities would be wiped away and annihilated. (Galileo, *Opere,* 6:347–48; trans. Drake, *Discoveries and Opinions,* 274)

And a little further on Galileo makes this concept more precise:

> To excite in us tastes, odors and sounds, I believe that nothing is required in external bodies except shapes, numbers, and slow or rapid movements. I think that if ears, tongues and noses were removed, shapes and numbers and motions would remain, but not odors or tastes or sounds. The latter, I believe, are no more than

names when separated from living beings. (Galileo, *Opere,* 6:350; trans. Drake, *Discoveries and Opinions,* 276–77)

These words of Galileo show clearly the way that modern science would proceed. In fact, it is only through translation into quantitative terms of the data of sense experience that it is possible to establish optics, acoustics, and all other parts of physics as sciences.

More in depth, these words imply an entirely new view of natural philosophy. In the place of the traditional one, with its interest for the qualitative properties of bodies, or "accidental forms," as the Aristotelians would say, as a way through which one can know their "essence," Galileo introduces the new natural philosophy of figures, numbers, and local motions of bodies. It is a philosophy that is based upon a "mechanistic" view of the material world, which is seen as an enormous mechanism whose macroscopic components come about through the aggregation of innumerable microscopic components, the "atoms" of Democritus and Lucretius.

Galileo certainly did not have the time (and perhaps not even the more speculative interest) required to come to a systematization of his philosophical ideas, which thus remain often in a fragmentary form, even though of great value and interest. In *The Assayer,* Galileo shows an extremely cautious attitude towards Copernicanism. The decree of 1616 always remained in force, and his adversaries were on the alert, ready to gather the least indication that Copernican convictions still kept hold on him. A proof of this was the dangerous allusion made by Grassi in the *Scale.* Galileo, therefore, never fails to profess repeatedly his submission to the decision of the Church on Copernicanism. But he nonetheless insists that one must prove also with "natural reason," when one can, the falseness of those propositions that are decided to be against Holy Scripture. The Church had not prohibited Copernicanism as a mathematical hypothesis. And Galileo does not fail to emphasize the fact that it is superior to all other theories proposed, even if one must then put it aside "for higher reasons":

If the movement attributed to the Earth, which I, as a pious and catholic person, consider most false and null, is suitable for explaining so many and diverse appearances that are observed in the heavenly bodies, I will not try to prove that it, so false, cannot even deceitfully respond to the appearances of the comets, if Sarsi does not descend to considerations more specific than those which he has produced up until now. (Galileo, *Opere,* 6:311)

Galileo obviously shows even more caution as to scriptural arguments. Sarsi had not held back in using scriptural quotations in support of his views. Not without a subtle irony, Galileo writes in this regard:

And since I could greatly fool myself in penetrating the true meaning of matters that by so great a margin go beyond the weakness of my brain, while leaving such determinations to the prudence of the masters of divinity, I will simply go on discussing those lower doctrines, declaring myself to be always prepared for every decree of superiors, despite whatever demonstration and experiment that would appear to be contrary to them. (Galileo, *Opere,* 6:360)

Urban VIII, no doubt, appreciated the prudence shown by Galileo and not less the vivacious style of the book, full of polemical irony, which spared nothing of the simplemindedness of Sarsi. Ciampoli had personally read to him some pages of it, and he informed Galileo about it, adding:

Here is much desired something else new from your talent, whence if you would resolve to have printed those concepts that until now remain in your mind, I am sure that they would be received most gratefully by Our Lord [the pope] who does not cease to admire your eminence in all matters and to retain intact for you that affection that he has had for you in times gone by. (Galileo, *Opere,* 13:146–47)

Grassi did not give in. His response was soon ready, but its publication was delayed more than had been expected, due to the fact that the book was being printed in Paris. Since *The Assayer* had been dedicated to the pope, who held it in great esteem, it was more prudent that a critical response to it not be published in Rome but in Paris, where its appearance would attract less attention. The title was *Comparison of the Weight [of the Arguments] of the Scale and of The Assayer,* and the pseudonym was again that of "Sarsi." Galileo read this work and annotated it with no less irony and severity than that which he had already employed with the *Scale.* But he did not think that it was worth the trouble to respond any further because, he noted, the author "was resolved to have the last word any way."

The tone of the *Comparison* was to all appearances well mannered, but its content was still polemical. And it did not lack insinuations, the most serious of which concerned the theological implications of the atomistic explanation of sense experience given by Galileo in *The Assayer.* Such an explanation, according to Sarsi, contradicted the Catholic doctrine of the Eucharist. According to the latter, the substance of bread and wine is transformed into the body and blood of Christ, but the "species" or sense qualities—namely, color, odor, and taste of bread and wine—are miraculously preserved. Now, this miracle was completely useless if these sense qualities were nothing else than "pure names," as asserted by Galileo. Grassi had stated beforehand that he had felt the need to make known in this way "a scruple that bothered me." In an annotation to this passage Galileo wrote:

> This scruple is left completely to you, because *The Assayer* is printed in Rome with the permission of superiors, and it is dedicated to the supreme head of the Church; it has been reviewed by those who are responsible for protecting the faith incorrupt and they, having approved, will have also considered how one might get rid of that scruple. (Galileo, *Opere,* 6:486, no. 149)

Galileo was nevertheless worried by this accusation of Grassi, and he asked Castelli to speak about it to Father Riccardi. The Dominican reassured him by stating that these opinions "were not otherwise against the Faith, since they were simply philosophical," and he offered to help him should there be the need.

Even before the publication of *The Assayer,* Galileo had thought about a new trip to Rome. The pretext was to pay homage to the new pontiff. But such an encounter with the pope would also have given Galileo the opportunity of discreetly sounding the possibility of a change in the Church's attitude towards Copernicanism. With his "admirer" as the head of the Church, and with Cesarini and Ciampoli in positions of great trust in the pontiff's court, it was "a marvelous combination of circumstances" that would never again present itself. This was what he wrote to Cesi in October 1623, asking for his advice on the matter. Cesi's answer was more than encouraging:

> Your coming here is necessary and it will very much please His Holiness who asked me if and when Your Lordship was coming; and I answered him that I was thinking that to you an hour seemed a thousand years, and I added that which seemed necessary with respect to the devotion of Your Lordship to him, and that I would soon bring him your work [*The Assayer*]; in a word, he showed that he loved you and esteemed you more than ever. (Galileo, *Opere,* 13:140)

Because of ill health, Galileo's trip to Rome had to be put off until the beginning of April. During a stop at Acquasparta, where Cesi was at that time, word arrived of the sudden death of Virginio Cesarini, an event that deprived Galileo of a stalwart supporter during his stay in Rome. Galileo arrived in Rome on April 23, and one day later he was received in audience by Urban VIII. The satisfaction with the first meeting, where the discussion must have been necessarily about matters in general, changed into a feeling of increasing uneasiness during the following five talks with the pope. Urban VIII had not responded in the way Galileo had hoped he would to his prudent soundings with respect to the issue of Copericanism. Convinced by now that "time, coolness and prudence" would be required, Galileo used his free time for various encounters with ecclesiastical authorities who could be of help to him. One among them was Cardinal Zollern, bishop of Osnabruck, who promised to help him. He did, in fact, do so on the occasion of his farewell visit to the pope, before his return to Germany. Galileo informed Cesi about it on June 8:

> Zollern left yesterday for Germany, and he told me that he had spoken with His Holiness on the matter of Copernicus, and how

the heretics are all of his opinion and hold it as most certain, and
that, therefore, one must go very circumspectly in coming to any
determination; to which His Holiness responded that the Holy
Church had not condemned it now as heretical, but only as temerarious, though it was not to be feared that there would be anyone to
demonstrate it as necessarily true. (Galileo, *Opere,* 13:182)

These last words show clearly what was, and will always remain, the
personal conviction of Urban VIII on the matter of astronomical systems. With all his admiration for Galileo, the pope was convinced that
no astronomer would ever be able to decipher the mystery of heavenly
motions, a mystery known only to God. It was a matter of a theologically
based skepticism as to the ability of human science to sound out the se-
crets of the universe. His skepticism contrasted in the sharpest way with
the conviction on which all of Galileo's scientific research was founded.

Without a doubt, Galileo himself had heard this statement directly
from the pope on the occasion of his audiences with him and perhaps
already at the time when Maffeo Barberini was a cardinal. In fact, we
know from Agostino Oreggi, papal theologian, of a conversation that
took place (we do not know when) between Barberini himself and Galileo. During it, the cardinal:

conceding him all that he had thought out, asked him at the end
whether God could not have the power and wisdom to dispose and
move in another way the orbs and the stars and all that is seen in
the sky and all that is said of the motions, order, location, distance,
and disposition of the stars. Because if God knew how, and had the
power to dispose all of this in another way than that which has been
thought, in such wise as to save all that has been said, we cannot limit
the divine power and wisdom to this way. Having heard this, that
most learned man remained silent. (Oreggi, *De Deo uno,* 194–95)

Galileo had to resign himself to no visible spectacular change in the
situation, a change that perhaps he had had the illusion of achieving
through his conversations with the pope. In the course of their conversations, it is possible that the question of Galileo's admonition by Bellarmine might have surfaced. His visit to the cardinal's residence, as

well as that of Commissary Segizzi, would not have escaped the attention of the Roman public. In fact, those visits must have been at the origin of the rumors, which soon enough started to spread in Rome, of a summons of Galileo by order of the Holy Office and of his abjuration of his Copernican convictions. The meeting of the Congregation of the Index, held for the preparation of the decree on Copernicanism, had taken place three days later. One can thus suppose that the cardinals who attended that meeting, including Maffeo Barberini, were already aware of that summoning of Galileo, and possibly put some questions on the matter to Bellarmine. The more so since they were aware that the decree, whose preparation had been entrusted to them, was aimed in particular to putting an end to the Copernican campaign of Galileo. We do not know, of course, how much Bellarmine would have been able to say on the matter. Since the cardinals attending the meeting were aware of the decision of Paul V that the Copernican theory was opposed to scripture, it is probable that he would have told them, at least, that the grand duke's mathematician had been in fact informed about that decision and that the Copernican theory could not be held. At any rate, he would not have said anything concerning the Segizzi intervention, which had taken place against instructions and which had, therefore, been illegal.

If such a hypothesis is in agreement with the facts, it is possible that at the questions of Urban VIII, if there were any, on this subject, Galileo had limited himself to answer what in fact he would declare during the 1633 trial, that is, that Bellarmine had told him that if defended in an absolute sense, the Copernican thesis could not be held, being contrary to scripture, but that it could be held *ex suppositione,* namely, as a purely mathematical hypothesis, as Bellarmine had already stated in his *Letter to Foscarini.*

Urban VIII could not but agree with Bellarmine on this point. As we know, according to him any attempt at demonstrating the Copernican theory as a possible physical explanation of the constitution of the world would end in failure, given the fact that the human intellect cannot possibly ever sound out the mysteries of the divine omnipotence. He thus must have ended by allowing Galileo to deal again with the Copernican issue as a purely mathematical hypothesis.

For Galileo, on the contrary, "hypothesis" meant an attempt at an explanation, not yet corroborated by sense experiences and certain demonstrations, which, however, could be so corroborated in the future, becoming thus a certain conclusion. As we will see, in writing the *Dialogue,* Galileo will try to play with the ambiguity of the concept of hypothesis. It was a risky game, which would convince nobody and, in the end, not even Urban VIII.

Another risky game, on Galileo's part, was perhaps that of not having spoken to the pope of a real "precept," received from Bellarmine, and certainly not of the one received by Commissary Segizzi eight years ago. His silence was most probably due to the fact that he felt that he was protected by the certificate of Bellarmine. Indeed, no mention of a precept was made in it, but only of the declaration of the pope with regard to the Copernican thesis, later published by the Congregation of the Index. Nor was the subsequent injunction by Segizzi in any way mentioned in that certificate. Naturally, Galileo did not know that the document in which that injunction was reported had been put in the files of the Holy Office. We will have the occasion to see in the next chapter how much the discovery of that document will influence in a negative way the whole of Galileo's trial.

In spite of this risky game, based at least in part on an overestimation of the strength of his relationship with the pope, Galileo was well aware of the necessity to be prudent in taking up again the discussion of the Copernican theory without allowing his adversaries to create an uproar against him. In agreement with the opinion of his friends, Galileo decided to proceed gradually. The first step was to compose a short essay in which the Copernican issue would be dealt with only in an indirect way.

The inspiration for his new writing was offered to him by the dissertation published eight years earlier by Francesco Ingoli, *A Disputation on the Location and Stability of the Earth against the System of Copernicus, to the Most Learned Mathematician Galileo Galilei.* Galileo had been prevented from responding to him because of the events of 1616. But now, after returning to Rome, he became aware of how much his silence had damaged him. Not knowing the reasons for his silence, many had interpreted them as due to the strength of Ingoli's arguments, which Galileo would not have been able to counter. With the approval of his Roman friends, Galileo thus decided to write an answer to Ingoli. From the reactions that it would arouse, Galileo would be able to decide whether or not to proceed further with the composition of the work for which he cared most.

Upon his return to Florence around the middle of June, he immediately went to work. The reply to Ingoli, in the form of a letter addressed to him, had already been completed towards the end of September 1624. Putting aside theological arguments, Galileo took up only the scientific aspect of the Copernican question. He stated right from the beginning that he had no intention "to rise again or support as true a proposition that had already been declared suspect and repugnant to a doctrine higher than physical and astronomical doctrines in dignity and authority." His intent, he specified, was to show to the Protestants that the exclusion of the Copernican system did not originate, on the part of Catholics, from ignorance of natural reasons, but "because of the reverence we have towards the writings of the Fathers, and because of our zeal in religion and faith" (Galileo, *Opere,* 6:510–11).

Those very "reasons, experiences, and observations" in favor of Copernicus were in fact piled up with an increasing force in the letter. Galileo could well tell Ingoli:

> Mr. Ingoli, if your philosophical sincerity and my old regard for you will allow me to say so, you should in all honesty have known that Nicolaus Copernicus had spent more years on these very difficult studies than you had spent days on them; so you should have been more careful and not let yourself be lightly persuaded that you could knock down such a man, especially with the sort of weapons you use, which are among the most common and trite objections advanced in this subject. (Galileo, *Opere,* 6:512; trans. Finocchiaro, *Galileo Affair,* 156–57)

Galileo had played a dangerous game. Would his statement of not being Copernican "for higher motives" be accepted in good faith in Rome? And how would his old adversaries and Ingoli himself have reacted? That Galileo was far from at peace in these matters we see from his recommendations of prudence to Cesare Marsili, his correspondent from Bologna, to whom, through Castelli, he had sent a copy of the *Letter to Ingoli,* for his comments.

The *Letter to Ingoli* arrived in Rome in October. Ciampoli reviewed it and decided to make some corrections, in order to prevent dangerous misunderstandings. But towards the end of March of the following year he had still not given the copy back to Guiducci, who had acted as intermediary between Ciampoli and Galileo. On April 18 Guiducci wrote to Galileo telling him that he had, with Cesi's advice, put off delivering the corrected copy to Ingoli. The reason, added Guiducci, was that some months before "a pious person" had proposed to the Holy Office that *The Assayer* be prohibited or at least corrected, since that book praised the teaching of Copernicus on the motion of the Earth.

Guiducci added that an (unnamed) cardinal had assigned the examination of *The Assayer* to a member of a religious order, Giovanni di Guevara. The latter had not only given approval to the book, but, Guiducci added, had also "put in writing some defenses to the effect that, even if that doctrine of motion were held, it did not seem to him to deserve condemnation; so the matter quieted down for the time being." But, Guevara, Guiducci continued, had gone off to France to accompany Cardinal Francesco Barberini. And so, given the absence of those two persons favorable to Galileo, Cesi was of the opinion that it was better not to run risks because

> in the letter to Ingoli Copernicus's opinion is explicitly defended, and though it is clearly stated that this opinion is found false by means of a superior light, nevertheless those who are not too sincere will not believe that and will be up in arms again. (Galileo, *Opere,* 13:265; trans. Finocchiaro, *Galileo Affair,* 205)

There was also the fact, Guiducci said again, that "we are opposed here by another powerful man, who once was one of your defenders." It is probable that he meant Cardinal Orsini. In a postscript to the letter he asserted, in fact: "The Lord Cardinal remains affectionate towards you, but Apelles has a great influence on His Most Illustrious Lordship" (Galileo, *Opere,* 13:266). Actually, Alessandro Orsini had some years before joined the Jesuit order. Thus Apelles (Scheiner) had become his fellow religious, and this could explain how the cardinal, under the latter's influence, could have become "contrary in this part [the Copernican opinion] whereas, at one time [in 1616] he took the lead in defending you."

In his 1987 book, *Galileo Heretic,* Pietro Redondi has asserted (rightly, in my opinion) that Guiducci, or his probable informer Cesi, had equivocated with regard to the content of the accusation made by that "pious person." The motion of which the latter spoke was not that of the Earth but rather that of the particles that, on the basis of the atomistic conception of matter adopted by Galileo, explained the origin of the sense qualities, as colors, odors, tastes, and so forth. As we have seen, Grassi, in the *Scale,* had seen this theory as opposed to the dogma of the Eucharist. A document discovered by Redondi shows an echo of such theological preoccupation and seems to be in fact that very one that was at the basis of Guiducci's information. This hypothesis makes much more plausible Guevara's favorable reaction, which on the contrary would be difficult to understand if it had concerned the Copernican theory, still affected by the decree of 1616. Redondi has further asserted that the author of the accusation was in fact Grassi. Such an identification, however, as well as his further claim that that accusation was the real cause of Galileo's trial, is lacking in convincing documentary support, and has not won the acceptance of the majority of Galilean scholars.

The final result of this warning from Guiducci was that Ingoli, in all probability, was never able to read the letter addressed to him. Neither was Grassi, who had got wind of its existence and had tried to get a copy from Guiducci—nor was anyone else, outside the circle of the most trusted friends of Galileo. As a consequence, the aim of sampling opinion that had been the reason of the letter's composition was lost in the end.

On the other hand, since December of the previous year, Ciampoli had informed Galileo that he had read to Urban VIII "a large part" of the *Letter to Ingoli* and that the pope had particularly liked some passages of it. But it is very probable that Ciampoli had chosen with dexterity those passages where there was no risk of arousing the pope's suspicions, and therefore one could not deduce too much from such a favorable reaction by him. Perhaps Galileo did not realize this and may have felt himself sufficiently protected by such a reaction "at the top." He therefore decided to go ahead with the work that was dearest to him and the first draft of which he had already completed several years before, the *Discourse on the Ebb and Flow of the Sea,* a work that would eventually be reworked as a dialogue and retitled as *Dialogue Concerning the Two Chief World Systems* (and which will hereafter be referred to as the *Dialogue*). In fact, he had already started dedicating his time to it, in parallel with the writing of the *Letter to Ingoli.*

At the beginning Galileo had hoped that he would be able to bring to quick completion this work of his. But he took sick in March 1626 and was forced to interrupt its composition. And during the next three years, progress was slow. In addition to his health, the situation was influenced by various other works and long-standing interests and research areas that were revivified in Galileo's mind as the *Dialogue* slowly went on developing. It was probably the massiveness of this book "on the system of the world," which Galileo had already announced in the *Starry Messenger,* that appeared now to him in all its vastness. The need to deepen so many topics and to refine so many arguments slowed him in the completion of the work. Above all else it was necessary to collect more data on the phenomenon of the tides, especially since difficulties had been raised on that subject, which was at the center of the book. An echo of those difficulties that Galileo had found himself facing during the composition of his work and that had been "considered by him always as almost insuperable," can be perceived in a letter written to Diodati in October 1629 and another two months later to Cesi.

While Galileo was slowly working on this new book, the Jesuit Scheiner was also carrying on in Rome (where he had resided since the year 1624) the composition of a new book on sunspots. It was finally published in 1630 and was entitled *Ursine Rose,* as a sign of deference towards Prince Paolo Orsini (brother of Cardinal Alessandro Orsini), whose press had taken care of the printing of the book and to whom the book was dedicated.

The *Ursine Rose* was a massive work of as many as 784 double-columned pages and collected the results of long years of observations.

Right from the beginning of the work, Scheiner took a harsh polemical attitude towards Galileo. In the *Discourse on Comets,* Galileo had made a sharp attack on Apelles, accusing him of plagiarism. And Scheiner had interpreted (perhaps wrongly) as addressed to him similar criticisms contained in *The Assayer.* Now it was his turn for "repaying [Galileo] in kind." His answer was indeed harsh and implacable. It took up the whole of book 1, a total of sixty-six pages. Scheiner claimed not only that his own discovery was independent from that of Galileo but also that he had the priority in the scientific study of the sunspots. In this he certainly held a good hand. In fact, his first letters to Welser had preceded by several months those of Galileo. Scheiner, however, went much further, launching a thorough attack against the scientific competence of Galileo. Here too he held a strong hand. Galileo, in fact, had not published anything more on the subject after his *Letters on the Sunspots* of 1613. Scheiner, on the contrary, had carried on and deepened his studies over many years, reaching important conclusions. And now he used his position of strength to "destroy" his adversary, by accusing him of crass ignorance of the most fundamental astronomical notions. Even Galileo's statement about the Sun's rotation on its own axis and that the spots were surface features of the Sun drew Scheiner's criticism as not being based on scientific reasons or even as being due to "chance."

The weighty and implacable polemic against Galileo with which this work is impregnated (even beyond book 1) was undoubtedly the source of the comments made about it, with a severity no less excessive and unjust, by Galileo's friends, who probably did not read beyond the first book. In fact, beginning with book 2 and continuing through most of book 4, Scheiner presented the results of his innumerable observations of sunspots carried out over a period of eighteen years, thus offering the most complete and most valuable treatise on solar physics of the epoch. By this time, the German Jesuit had become convinced of the rotation of the Sun around its own axis and also of the fact that spots were surface features of the Sun. He had, furthermore, determined that their trajectories were curved, and he even presented the dates of the maxima and minima of the curvature with a truly noteworthy precision, given the instruments of that time.

The conclusion of Scheiner on the sunspots as a solar phenomenon put him in direct contrast with the Aristotelian affirmation of the incorruptibility of heavenly bodies. Scheiner therefore dedicated the final part of his book to the criticism of Aristotle's position. And this undoubtedly constitutes a further merit of the *Ursine Rose*. It is also to be noticed that the Jesuits (among whom Grassi) in charge of the review of the book for permission to print, gave a very positive appraisal of it. This can be interpreted as a vindication of the freedom of scientific research against the constraints of "fidelity" to Aristotelianism.

It appears that Galileo was in a position to have a copy of the *Ursine Rose* only towards the end of 1631. Following the advice of his friends, he did not answer Scheiner, content with what he had already written about him in the *Dialogue*. In fact, the latter had already been finished at the beginning of 1630, and now came the question of its publication. Answering a previous request of Galileo, Cesi and the other Roman friends agreed that it would be advantageous that this new work also be published in Rome.

Once again, the events seemed to favor Galileo. As of June 1629, Father Riccardi, who had given such a favorable judgment on *The Assayer,* had been appointed as master of the Sacred Palace. Given this charge, it would be he who had to give the permission for the printing of the *Dialogue.* Another favorable circumstance was the presence in Rome of Castelli, who had been called there by Urban VIII to be a tutor of his nephew Taddeo Barberini and who had afterwards been appointed lecturer of mathematics at the Roman university La Sapienza.

Castelli had had already the opportunity to speak with Riccardi about the *Dialogue* and assured Galileo "that for his part he was certain that things will be going well." Even Ciampoli, Castelli added, had shown himself cautiously confident. He could not promise anything for certain, but Galileo's coming to Rome would make it possible to overcome whatever difficulty might arise.

In fact, Galileo arrived in Rome on May 3, 1630, where he was, as usual, the guest of the ambassador of the grand duke, who was, since 1621, Francesco Niccolini. And Galileo was received by him with warmth, which was in clear contrast with the antipathy shown towards him by Niccolini's predecessor, Guicciardini. And an equal warmth and sympathy was shown to him by the ambassador's wife, Caterina Riccardi, a relative of Father Riccardi.

Everything, therefore, seemed to promise a happy outcome of this trip to Rome, which Galileo was anticipating would provide the crowning of his struggles, his efforts, and his hopes. On May 18 he was received by Urban VIII. On the basis of what the pope himself stated two years later, we may surmise that during the conversation Galileo must have manifested his intention to print his new book. Urban VIII was certainly very explicit in repeating his own position with respect to the problem of Copernicanism. Still, from the pope's attitude Galileo must have concluded that he was not opposed in principle to the publication of the *Dialogue*.

While Galileo was in Rome, an event occurred that could have had the direst of consequences for him. During the first half of May news spread of a horoscope, attributed to the Vallombrosian Father Orazio Morandi, which foretold the approaching death of Urban VIII and of his nephew Taddeo Barberini. Now Morandi was a friend of Galileo. Thus some of Galileo's enemies tried to implicate him also in the affair. And there were some who even went so far as to make Galileo the author of the horoscope.

Urban VIII was very superstitious. The horoscope, moreover, threatened to create a "vacant See" psychology and consequently a situation of political vacuum. And this the pope could certainly not allow to happen. He therefore had Morandi imprisoned shortly before Galileo's departure from Rome. The unfortunate father died in prison towards the end of the same year before the beginning of his trial.

Galileo did not find it difficult to persuade Cardinal Francesco Barberini and, through him, Urban VIII himself that he had not played any part at all in that affair. But the incident showed how strong and tenacious was the hatred harbored against him by his enemies. Knowing that the pope had great esteem for him, it was precisely this esteem that they had sought to ruin by leaning upon the susceptibility and impressionability of Urban VIII.

Another person whom they tried to involve in the affair of the horoscope was Tommaso Campanella. He had been transferred from the Neapolitan prison to that of the Roman Inquisition in 1626, thanks to Urban VIII's intervention, and had been freed in 1629. At that moment he enjoyed the pope's favor. Shortly after his arrival in Rome, Galileo had delivered the manuscript of the *Dialogue* to Father Riccardi, in order to obtain from him the *imprimatur,* the permission to print. Without doubt Riccardi was up to date on Urban VIII's ideas with respect to Copernicanism. Moreover, the decree of the Index of 1616 was still in force. Riccardi knew, therefore, that he had to be prudent, despite the signs of esteem that the pope continued to show to Galileo. The reading of the manuscript must thus have brought him more than one moment of apprehension. Despite the protests, repeated here and there in the book, that he had left the question open, Galileo's attitude was clearly in favor of Copernicus. Riccardi, therefore, decided that the book had to be revised. It should be supplied with a preface and a conclusion in which it would be made sufficiently clear that one was dealing with hypothetical reasoning. The rest of the book was also to be adapted "in such wise as to show that the Holy Congregation [of the Index] in disapproving Copernicus had acted in an entirely reasonable way" (Galileo, *Opere,* 19:325), as Riccardi himself will write at the moment of Galileo's trial.

The job of reviewing the manuscript was given to another Dominican, Raffaele Visconti, a professor of mathematics and friend of Riccardi. He made various corrections, but for the rest he approved the book and declared that he was ready to give a testimonial in its favor. Riccardi, however, was not of the same opinion and decided to reexamine the whole book by himself. Galileo complained about this second review, which could delay the printing. Riccardi, therefore, agreed to give the *imprimatur* for Rome so that Galileo could start the negotiations with the publishers. The master of the Sacred Palace would then have the task of delivering to the latter the pages of the manuscript as he had reviewed them.

Finally, on June 16 Visconti was able to give Galileo the much sought-for news:

> The Father Master [Riccardi] says that he likes the work and that tomorrow morning he will speak with the pope for the frontispiece of the work and, for the rest, after arranging a few little things, like those which we settled together, he will give you the book. (Galileo, *Opere,* 14:120)

We do not know anything about the outcome of that audience. It is probable that Riccardi limited himself to dealing with the question of the title of the book, with some vaguely reassuring phrase about its content. And it must have been on this occasion that Urban VIII requested that the title of the book be changed from the original one, *Dialogue on the Ebb and Flow of the Sea,* to that of *Dialogue Concerning the Two Chief World Systems* or something similar. Obviously, Urban VIII did not want it to be seen right from the beginning that the book was based principally on the explanation of the phenomenon of the tides as due to the Earth's motion. According to what the pope would affirm later, he did not give the "permission" for the printing of the book on that occasion. Nor, it seems, did he give it anytime later.

Galileo had heard from Visconti (to whom Riccardi must have spoken about the audience with the pope) that Urban VIII had shown "annoyance" concerning the demonstration of the Earth's motion through the phenomenon of the tides. Visconti had promised to speak personally to the pope, with the hope to "deliver him from that annoyance." But he was not able to do so. Having been involved himself in the Morandi affair, Visconti was subsequently exiled from Rome.

If Urban VIII did not give his assent to the printing on that occasion, how was it possible for Riccardi to give the *imprimatur* by himself? According to what Gianfrancesco Buonamici affirmed immediately after Galileo's condemnation, it would have been Ciampoli making Riccardi believe that he himself had received from the pope the assent for the permission to print.

On June 26 Galileo left Rome, convinced that he had accomplished the purpose of his trip. Even Ambassador Niccolini, writing to the secretary of state of the grand duke, Andrea Cioli, showed himself convinced of that success. And he reported that the leave-taking from the pope had been most cordial and that Galileo had been invited to dinner by Cardinal Francesco Barberini.

Obviously those signs of benevolence towards Galileo, shown by the pope and by the Cardinal-Nephew Francesco Barberini, were to be attributed to the fact that no one (except Riccardi and Visconti) was aware of the true content of the book. Upon his return to Florence, Galileo intended to compose the index, the dedicatory letter, and a few other things that were missing, and then to send the manuscript immediately to Cesi, who had decided to print this book under the auspices of the Lincean Academy. But at that crucial moment, Cesi suddenly died on August 1. With his death, Galileo lost one of his most genuine and valuable friends. And this event will have a deeply negative influence on the whole further course of events.

Cesi had left no will, which provoked a situation of crisis at the Lincean Academy. On the other hand, Castelli wrote to Galileo on August 24 that "for many most weighty reasons that I do not wish to commit to paper at this time" he suggested to provide for the printing of the book in Florence and "to do this as soon as possible." We do not know to what Castelli was alluding. Galileo's enemies had probably continued to agitate, in spite of the failure of their attempt to involve Galileo in the Morandi case, the more so now that some news had inevitably leaked concerning the new book of Galileo. One cannot absolutely exclude that there was also among them a Jesuit, such as Scheiner. In fact, the latter will be seen, after Galileo's condemnation, as the principal one responsible for its orchestration. I shall return in the next chapter to the question of his responsibility and more in general of that of the Jesuit Order in this matter.

Galileo ended by following Castelli's advice. Now, however, the *imprimatur* had to be granted not in Rome but in Florence. Having been consulted on the matter, Riccardi gave as a condition of his consent that Galileo send him a copy of the *Dialogue,* so that he might be able to adjust, together with Ciampoli, "some small matters in the introduction and in the book itself" (Galileo, *Opere,* 14:150).

But just then the plague was rampant in Italy. It had spread rapidly from the north towards Tuscany and was now menacing the Church States. Rigorous quarantine measures (even for books and manuscripts) had been imposed in the latter, and Galileo, in order to avoid further loss of time, requested permission to do the final revision of the book in Florence. He would have sent to Rome both the preface and the final part for the corrections that Riccardi wanted done.

After long and extensive negotiations, which were protracted up to May of the following year, Riccardi in the end decided to send his instructions to the Florentine Inquisitor Clemente Egidi, whom he asked to determine by himself whether or not to give the *imprimatur.* According to these instructions, Galileo had to insert at the beginning of his book a preface written by Riccardi himself, in which it was stated that the Copernican theory was treated in it as a purely mathematical hypothesis. At the conclusion of the book, Galileo had to introduce the theological argument of the pope, reaffirming the impossibility, for the human mind, to penetrate the mystery of the real constitution of the world, known to God alone. These instructions make clear the fundamental misunderstanding at the basis of the permission given to the publication of Galileo's work. If, in fact, this was conceived as a purely mathematical hypothesis, the *medicina finale* (the final medicine) of the pope's argument was unnecessary. Was Riccardi himself unaware of such contradiction? Having personally examined the book, he must have been well aware of its real aim and content, and that was undoubtedly at the origin of his long hesitations at granting the *imprimatur.* Maybe he felt somehow relieved, at the end, by the fact that he had transferred upon the poor Egidi the final responsibility of the whole affair.

At any rate, the long-desired permission had thus virtually arrived, and Galileo was finally ready to see to the beginning of the printing. The work, however, was slow, given also the great number (one thousand) of copies to be printed. And thus it was only on February 21, 1632, that the typographer, Landini, was able to announce to Cesare Marsili in Bologna that the printing had been completed.

The *Dialogue* carried the ecclesiastical printing permission of the vice-regent of Rome, of the master of the Sacred Palace, of the vicar general of Florence, and of the Florentine inquisitor, in addition to that of the government of the grand duke. And the printing date was the year 1632. As is evident, Galileo had wanted to insert all the possible ecclesiastical permissions. The two Roman permissions, including that of Riccardi, however, were out of place, since the book was printed in Florence. And this will, in fact, be one of the charges against Galileo at the time of his trial.

The title had been composed according to the wish of Urban VIII. In full, it read *Dialogue of Galileo Galilei, Lankan, Special Mathematician of the University of Pisa and Philosopher and Chief Mathematician of the Most Serene Grand Duke of Tuscany. Where, in the Meetings of Four Days, There Is Discussion Concerning the Two Chief Systems of the World.*

I will limit myself here to a brief comprehensive view of this work, whose content was summarized at its beginning by Galileo himself as follows:

> Three principal headings are treated. First, I shall try to show that all experiments practicable upon the Earth are insufficient measures for providing its mobility, since they are indifferently adaptable to an Earth in motion or at rest. I hope in so doing to reveal many observations unknown to the ancients. Secondly, the celestial phenomena will be examined, strengthening the Copernican hypothesis until it might seem that this must triumph absolutely. Here new reflections are adjoined, which might be used in order to simplify astronomy, though not because of any necessity imposed by nature. In the third place, I shall propose an ingenuous speculation. It happens that long ago, I said that the unsolved problem of the ocean tides might receive some light from assuming the motion of the Earth. This assertion of mine, passing by word of mouth, found loving fathers who adopted it as a child of their own ingenuity. Now, so that no stranger may ever appear who, arming himself with our weapons, shall charge us with want of attention to such an important matter, I have thought it good to reveal those probabilities which might render this plausible, given that the Earth moves.
>
> I hope that from these considerations the world will come to know that if other nations have navigated more, we have not theorized less. It is not for failing to take count of what others have thought that we have yielded to asserting that the Earth is motionless, and holding the contrary to be a mere mathematical caprice, but (if for nothing else) for those reasons that are supplied by piety, religion, the knowledge of Divine Omnipotence, and a consciousness of the limitations of the human mind. (Galileo, *Opere,* 7:30; trans. Drake, *Dialogue,* 6)

As we see, Galileo is very careful. He purports to hold the hypothesis of the Earth's motion as a "mere mathematical caprice" and concludes with an allusion to the argument of Urban VIII on the incapacity of the human intellect to decipher the mysteries of the universe. In reality, the *Dialogue* will develop in a totally different way.

Galileo then proceeds to introduce the three interlocutors of the *Dialogue*. Two of them were close friends of Galileo, by now deceased, Salviati and Sagredo. The third one, a Peripatetic (that is, Aristotelian) philosopher, to whom Galileo gives the name of the famous sixth-century commentator of Aristotle, Simplicius. It is however certain that Galileo chose that name because of its assonance with the Italian word *semplicione* (simpleton).

Salviati presents, as a specialist, the Copernican point of view. Sagredo represents an educated layman, full of interest for the new doctrines. Simplicius, on the contrary, plays the role of an Aristotelian philosopher. He too is an educated person in the manner of the dominant university culture of those times. And his arguments, one after the other, fall to pieces before the close critique of Salviati, so that ultimately Simplicius comes across as a simpleton who believes blindly in a natural philosophy no longer supportable.

The *Dialogue* takes place imaginatively on an undetermined date in Venice, at the Sagredo Palace on the Grand Canal, and it is divided into four "days." The first day is dedicated to a thorough critique of the fundamental presuppositions of the Aristotelian-Ptolemaic vision of the world. First of all, on a theoretical level, Salviati denies the Aristotelian distinction between two fundamental types of "natural movements," those rectilinear, in the case of the Earthly bodies, and circular, in the case of the heavenly ones. In an ordered universe, affirms Salviati, the only natural movements are the circular ones.

This affirmation nullifies not only the Aristotelian argument of the centrality and immobility of the Earth, but also the distinction between "ungenerated" and "incorruptible" heavenly bodies and the "generated" and "corruptible" terrestrial ones. Salviati states that all bodies in the universe are equally subject to change. Simplicius, naturally, protests: "This way of philosophizing tends to subvert all natural philosophy, and to disorder and set in confusion heaven and Earth and the whole universe." But Salviati answers him with courteous irony:

Do not worry yourself about heaven and Earth, nor fear either their subversion or the ruin of philosophy. As to heaven, it is in vain that you fear for that which you yourself hold to be inalterable and invariant. As for the Earth, we seek rather to ennoble and perfect it when we strive to make it like the celestial bodies, and, as it were, place it in heaven, from which your philosophers have banished it. (Galileo, *Opere,* 7:62; trans. Drake, *Dialogue,* 37)

Moving then to sense experience, Salviati denies Simplicius's claim that the immutability of the heavenly bodies is confirmed by the fact that no change has ever been noted in them. And he does it by bringing to bear the heap of new facts that have come to light from the recent telescopic discoveries. And he concludes: "I declare that we do have in our age new events and observations such that if Aristotle were now alive, I have no doubt that he would change his position" (Galileo, *Opere,* 7:75; trans. Drake, *Dialogue,* 50).

The second day has as its central theme that of the daily motion of the heavenly bodies. This motion, observes Salviati, could be explained either by the rotational motion of the Earth or by the motion of the rest of the world, excluding the Earth. Against this last opinion (which is the one of Aristotle and Ptolemy) Salviati shows with various reasons how the opinion of Copernicus of the motion of the Earth alone is more probable. This refutation of the Aristotelian-Ptolemaic system is one of the strongest points of the *Dialogue* and provides the occasion for Galileo to expound on the principles of kinematics that he had gone on formulating during the years in Padua.

Of particular importance is the refutation of the "classic" argument against the motion of the Earth. According to it, if the Earth really did move, then the effect of this should be seen in the deviation to the west of a mass falling from the top of a tower. This is exactly the same phenomenon (always according to the Aristotelians) that is visible in the displacement towards the stern of a stone which is let fall from the top of a ship's mast, when the ship is in motion. Salviati answers by denying the existence of the latter phenomenon. That is, the falling mass follows the same law both when the "reference system" is at rest and when it is moving. And so he can conclude: "Therefore, the same cause holding good on the Earth as on the ship, nothing can be inferred about the Earth's motion or rest from the stone falling always perpendicularly to the foot of the tower" (Galileo, *Opere,* 7:370; trans.

Drake, *Dialogue,* 342). We have here the formulation of the principle that today is rightly called the "principle of Galilean relativity."

In the third day, the debate moves on the question of the motion of revolution of the Earth around the Sun. The two most serious objections against it, from the scientific point of view, are the invariability during the year of the apparent brightness of the fixed stars and of their position in the heavenly vault. These objections (resumed in the *Mathematical Discussions* of Scheiner, amply quoted by Simplicius) are answered by Salviati (following Copernicus) by stating that given the enormous distances of the fixed stars from the Earth, such phenomena are not observable. Galileo was aware of the difficulties proposed by Tycho Brahe against this idea of the enormous dimensions of the universe (see chapter 1 of this book). And he gives the answer by stressing the fact of the reduction of the angular size of the fixed stars when it is calculated with the help of a telescope, compared with that calculated by Brahe. For example, in the case of a sixth magnitude star (supposed as intrinsically as bright as the Sun) its distance calculated from the telescopic measure of its angular size would be such as to make it practically impossible to detect a parallax.

This discussion on the enormous dimensions of the universe suggests to Salviati the question whether the universe is finite or infinite. Following Copernicus's prudent attitude on the matter, Galileo leaves it open. It is quite probable that, during his stay in Padua, Galileo had heard about Giordano Bruno's conception of an infinite universe. And naturally he had known about his condemnation and subsequent burning at the stake. One reason more, obviously, to be prudent. On the other hand, continued Salviati, there are enormous difficulties for the Ptolemaic system. And the most important thing is that the solution of them is found precisely by adopting the Copernican point of view. Such is above all the case for the explanation of the irregular motion of the planets. To this end, noticed Salviati: "Ptolemy introduces vast epicycles, adapting them one by one to each planet, with certain rules about incongruous motions, all of which can be done away with by the very simple motion of the Earth" (Galileo, *Opere,* 7:370). And Salviati shows how this comes about whenever one adopts the hypothesis of the Earth's motion around the Sun.

As a matter of fact, as we know, Copernicus had kept the idea of the circular motions of the heavenly bodies. He had thus been obliged to keep a certain number of epicycles in order to obtain a satisfactory agreement between the theoretical calculations and the observational data. A decisive simplification had been obtained with Kepler's theory of the elliptical orbits of the planets. But Galileo had continued to follow Copernicus even in this matter, based on his principle that circular motion is the only natural movement. Thus, Salviati's claim was only in part in agreement with the facts.

The discoveries with the telescope, too, are brought forward by Salviati. If they do not prove the Copernican view, "they absolutely favor it, and greatly." In particular, the discovery of the satellites of Jupiter eliminated the "apparent absurdity" that none of the planets had a satellite, while the Earth had one (the Moon). The discussion then turns on the sunspots. In a clear polemic with Scheiner, Salviati states: "The original discoverer and the observer of the spots (as indeed of all other novelties in the skies) was our Lankan Academician" (in the *Dialogue,* Galileo is not mentioned directly by name, but only with the latter title).

Salviati emphasizes in the first place the value to be drawn from the phenomenon of the spots, considered as part of the Sun's surface, in order to disprove the incorruptibility of the heavenly bodies. But this phenomenon, he adds, offers also a possible confirmation of the Earth's motion. In this regard, Salviati cites the phenomenon of the curvature of the trajectories described by the spots, as seen from the Earth, due to the rotation of the Sun around its axis, and of the variation (in a period of one year) of the form and inclination of that curvature. As we have seen, this phenomenon had been accurately observed by Scheiner and described in the *Ursine Rose.* It seems quite probable that Galileo—after reading that book—had time to introduce a mention of the phenomenon during the printing of the *Dialogue* (without mentioning its source: later on Scheiner would accuse him of plagiarism). But Galileo differs from Scheiner by giving an explanation according to the Copernican system. In substance he states, through Salviati, that the phenomenon can be easily explained by introducing two suppositions. The first one is that of the daily and annual motion of the Earth. The second one is that of a rotational motion of the Sun on its axis (in less than one month), which is inclined with respect to the ecliptic. On the contrary, in the case of the geostatic theory of Ptolemy, one would have to attribute up to four different motions to the Sun, partly incongruous among themselves. In conclusion, Salviati claims, the Copernican explanation appears more likely because it is simpler.

The fourth day is the culminating point of the *Dialogue,* completely focused as it is on the phenomenon of the ebb and flow of the tides, which, as we know, was for Galileo the strongest argument in favor of the Copernican system. Before all else, Salviati acknowledges that, on the basis of the principle already explained in the second day, terrestrial phenomena are not capable of confirming the motion or the stability of the Earth. But he adds:

> It is only in the element of water (as something which is very vast and is not joined and linked with the terrestrial globe as are all its solid parts, but is rather, because of its fluidity, free and separate and a law unto itself), that we may recognize some trace and indication of the Earth's behavior in regard to motion and rest. (Galileo, *Opere,* 7:442–43; trans. Drake, *Dialogue,* 417)

Right after this, Salviati offers what were the fruits of lengthy reflections.

> After having many times examined for myself the effects and the events, partly seen and partly heard from other people, which are observed in the movements of the water; after, moreover, having read and listened to the great follies which many people have put forth as causes for these events, I have arrived at two conclusions which were not lightly to be drawn and granted. Certain necessary assumptions having been made, these are that if the terrestrial globe were immovable, the ebb and flow of the oceans could not occur naturally; and that when we confer upon the globe the movements just assigned to it, the seas are necessarily subjected to an ebb and flow agreeing in all respects with what is to be observed in them. (Galileo, *Opere,* 7:443; trans. Drake, *Dialogue,* 417)

As is evident from this quotation, Galileo was certainly not ignorant of the traditional explanation of the tides phenomenon, especially of the one based on the influence of the Moon, which Simplicius at this point makes as a counter-proposal to Salviati. The latter puts aside these explanations, qualifying them as "poetical." And he includes among them even that of Kepler, wondering why such a man as he, "of open and acute mind and who had at his fingertips the motions attributed to the Earth had nevertheless lent his ear and his assent to the Moon's dominion over the waters, to occult properties and to such puerilities" (Galileo, *Opere,* 7:486; trans. Drake, *Dialogue,* 462). Here, as in other instances, Galileo shows himself unjust towards Kepler, who in the preface of the *New Astronomy* had sketched a theory of the tides based upon an attractive force acting between the Earth and the Moon, which surely was not "a puerility."

Salviati then goes on with a long exposition of the argument of Galileo. This "proof" of Galileo is today generally considered as completely erroneous. In fact, there is an effect on the tides of the Earth's rotation, and in this sense Galileo's intuition was substantially correct. But the error is on the quantitative level. The size of this effect is, in fact, very small and is, therefore, completely masked by the principal cause of the tides, namely the attraction of the Moon, combined with that of the Sun. And even if the effect of the Earth's motion had been more important than in fact it is, it would have shown itself in a sense different from that of the actual motion of the tides.

Now the *Dialogue* is winding towards its end, and as usual Sagredo is given the job of summarizing the fruit of the long discussions of the four days.

> In the conversations of these four days, we have, then, strong evidences in favor of the Copernican system, among which three have been shown to be very convincing, those taken from the stoppings and retrograde motions of the planets, and their approaches towards and recessions from the Earth; second, from the revolution of the Sun upon itself, and from what is to be observed in the Sunspots; and third, from the ebbing and flowing of the ocean tides. (Galileo, *Opere,* 7:487; trans. Drake, *Dialogue,* 462)

But the moment has come for Salviati to "remove the mask" of a follower of Copernicus, as he had promised at the beginning of the *Dialogue.* And he does it with these words:

> Now, since it is time to put an end to our discourses, it remains for me to beg you that if later, in going over the things that I have brought out, you should meet with any difficulty or any question not completely resolved, you will excuse my deficiency because of the novelty of the concept and the limitations of my abilities; then because of the magnitude of the subject; and finally because I do not claim and have not claimed from others the assent which I myself do not give to this invention, which may very easily turn out to be a most foolish hallucination and a majestic paradox.
>
> To you, Sagredo, though during my arguments you have shown yourself satisfied with some of my ideas, and have approved them highly, I say that I take this to have arisen partly from their novelty rather than from their certainty, and even more from your courteous wish to afford me by your assent that pleasure which one naturally feels at the approbation and praise of what is one's own. (Galileo, *Opere,* 7:487–88; trans. Drake, *Dialogue,* 463)

This was no doubt the original ending of the *Dialogue.* Affirmations as such could certainly not convince the readers of the book. Even less convincing was the appearance, as a sort of *deus ex machina,* of Urban VIII's argument, which Galileo had to insert here following the orders by Riccardi. After Salviati's words, the only person of whom Galileo had the possibility to put it in the mouth was, very unfortunately, Simplicius, the "simpleton":

> As to the discourses we have held, and especially this last one concerning the reasons for the ebbing and flowing of the ocean, I am really not entirely convinced; but from such feeble ideas of the matter as I have formed, I admit that your thoughts seem to me more ingenious than many others I have heard. I do not therefore consider them true and conclusive; instead, keeping always before my mind's eye a most solid doctrine that I once heard from a most eminent and learned person, and before which one must fall silent, I

know that if asked whether God in His infinite power and wisdom could have conferred upon the watery element its observed reciprocating motion using some other means than moving its containing vessels, both of you would reply that He could have, and that He would have known how to do this in many ways which are unthinkable to our minds. From this I forthwith conclude that, this being so, it would be excessive boldness for anyone to limit and restrict the Divine power and wisdom to some particular fancy of his own. (Galileo, *Opere,* 7:488; trans. Drake, *Dialogue,* 464)

For once, Salviati finds himself in agreement with Simplicius:

An admirable and angelic doctrine, and well in accord with another one, also Divine, which, while it grants to us the right to argue about the constitution of the universe (perhaps in order that the working of the human mind shall not be curtailed or made lazy) adds that we cannot discover the work of His hands. Let us, then, exercise these activities permitted to us and ordained by God, that we may recognize and thereby so much the more admire His greatness, however much less fit we may find ourselves to penetrate the profound depths of His infinite wisdom. (Galileo, *Opere,* 7:489; trans. Drake, *Galileo Affair,* 464)

With these words, the *Dialogue* practically comes to an end. It is a work that is not, nor was it intended to be, a dry treatise of astronomy and natural philosophy, but rather a polemical and at the same time a pedagogical writing, in support of the Copernican conception. In writing it, Galileo remembered the long years of battles with the Aristotelians, their unshakable opposition to new ideas and to new discoveries. He knew well that it presented an entire worldview, one in opposition to the *Physics* and *On the Heavens* of Aristotle, which had to be demolished so as to prepare the way for a recognition of Copernicanism by educated persons (including ecclesiastics) in Italy and all of Europe.

To be sure, Galileo did not offer, in the *Dialogue,* rigorously valid and "decisive" proofs in favor of the Earth's motion. And he himself must have been aware of this. He had, however, eliminated with a perfect scientific rigor many of the obstacles that hindered the acceptance of Copernicanism, including those deriving from the convictions of "common sense" that still had so much weight, even in his days, in the minds of men like Bellarmine. This scientific rigor of Galileo was based on results already obtained in the field of astronomical observations, as well as in that of the new "natural philosophy," on whose foundations he had worked since his years in Padua and which was now for the first time presented to the educated public. In doing so, Galileo had without a doubt made a decisive contribution to the triumph of Copernicanism.

But the *Dialogue* is also a profoundly human work, full of the richness of Galileo's personality, and not less of the drama of his life. It is an interior drama that is transparent through the pages of this book, before its exterior manifestation in the trial and the condemnation of its author. There are the repeated protests of uncertainty, of doubt, precisely at the moment when the conclusion in favor of Copernicus seems to be the most convincing; there are the words such as "hallucination," "fantasy," "paradox," which suddenly come forth upon a tightly-knit rigorous reasoning; there is especially the claim of Salviati that his Copernicanism is no more than a "mask," and that his true conviction is the one shown at the end of the *Dialogue* with his unconditional assent to the argument of Simplicius. All of this has scandalized many biographers of Galileo, so that they have accused him of being two-faced, of playing around disloyally and of being afraid. But if Galileo had really wished to play at being two-faced, to seek protection, he should have covered his hand better. In fact, the very evidence from Galileo's game, which would not go unnoticed by his judges, is the indication of a temperament too sincere to be able to successfully play games.

Sixteen years before, Galileo himself had written in the *Letter to the Grand Duchess Christina*:

> To command that the very professors of astronomy themselves see to the refutation of their own observations and proofs as mere fallacies and sophisms is to enjoin something that lies beyond any possibility of accomplishment. (Galileo, *Opere,* 5:325; trans. Drake, *Discoveries and Opinions,* 193)

The *Dialogue* is the clearest proof of that impossibility.

The Trial and Condemnation
of Galileo

As soon as it was published, the *Dialogue* began to be spread about in Italy and abroad, thanks in no small part to the numerous copies that Galileo had sent to friends and influential people. Father Riccardi also received a copy, the one that was sent by the inquisitor of Florence to the Holy Office and then redirected, as usual, to him. In the letter of reply to the inquisitor (March 6, 1632), Riccardi acknowledged receipt of the book without any comment on the matter.

Galileo's friends welcomed the *Dialogue* with enthusiasm, although frank reservations on the argument from the tides were not lacking. Of course, among the most enthusiastic was Castelli, who had been able to borrow from Cardinal Francesco Barberini a copy of the book, namely, one of those left unbound (in order to avoid the severe quarantine measures taken because of the raging plague) that had arrived in Rome at the end of April. In the letter sent to Galileo on May 29, Castelli manifested his great admiration for the new work. At the end he added:

> Monsignor Ciampoli continues to carry out his assignment, and there is no news other than the previous; and monsignor carries

on splendidly, with due esteem for the masters, and laughing to himself at the things of the world, as they deserve. (Galileo, *Opere*, 14:358)

This additional remark was intended as an answer to a question addressed to him by Galileo, as to rumors that were going about in Florence that Ciampoli had fallen out of favor with Urban VIII. In spite of Castelli's reassuring words, Ciampoli in fact had lost the pope's confidence since the month before. One of the important reasons is perhaps that he had aligned himself with the group of the Spanish-leaning cardinals at a political moment extremely difficult for Urban VIII.

From the beginning of his pontificate, the pope had supported France with the aim of counterbalancing the weight of a Hapsburg hegemony, which could have been the fruit of an understanding between Spain and the German Empire. As a result, he had now found himself, in one of the most dramatic phases of the Thirty Years War, in favor of an agreement between the King of France, Louis XIII, the Duke of Bavaria (representing the neutral Catholic League of Germany), and the Protestant Gustavus Adolphus, the King of Sweden, at that time at the apex of his military successes against the German Empire. This was considered by the partisans of Spain and of the empire as a betrayal of the Catholic cause.

In March 1632 the tension between the Hapsburg party and the pope became manifest in a most clamorous way. During a consistory, Cardinal Gaspare Borgia, who carried out the functions of ambassador of Spain, openly attacked Urban VIII, with accusations of favoring the cause of the heretics, and invited him to show that same "apostolic zeal" which had characterized his "more pious and more glorious" predecessors.

A different reason for why Ciampoli had fallen out of favor with Urban VIII is that given by Ambassador Niccolini. The pope had written a pastoral letter in Latin and distributed copies of it as a first showing among the cardinals and the members of the diplomatic corp. It seems that Ciampoli criticized the Latin style of the pope, composing a more elegant letter and showing it to several people. This would have deeply wounded Urban VIII.

The situation had been made even thornier for the pope because of accusations of nepotism and of earthly ambitions, made against him by certain Romans. Urban VIII sensed that he would have to intervene before it was too late. And that led to a hardening of his position, a susceptibility born of exasperation and suspicion, which characterizes this period of his pontificate. Ciampoli was one of the first victims of this new posture of the pope. We will see right away its equally negative influence with respect to Galileo.

Besides the praise and admiration of Galileo's friends, the *Dialogue* had certainly begun also to awaken the suspicion and the indignation of his enemies. This, however, must have occurred after some delay, since only a few copies of the book had reached Rome by the end of June, given the continuation of quarantine measures because of the still raging plague. And it is likely, even though it cannot be strictly proven, that Galileo's adversaries had already succeeded around the beginning of July at having their voice heard by Urban VIII. Whatever the case, we know, through confidential information from Riccardi to Magalotti at the beginning of August, that the *Dialogue* had come (probably around the middle of that month) into the pope's hands and that he had found it not altogether satisfactory (*manchevole*). This is clearly indicated in the letter sent by Riccardi to the Florentine inquisitor Clemente Egidi:

> There has arrived here the book of Mr. Galileo, and there are things that are not acceptable, and the masters wish at any rate that it be revised. Meanwhile it is the order of Our Lordship (but no more than my name is to be mentioned) that the book be withheld and that it not be sent here without there having been sent from here that which is to be corrected, nor should it be sent to other places. Please, have an understanding about it, Your most Reverend Paternity, with the Illustrious Monsignor Nuncius; and working in a pleasant way, see that everything succeeds efficaciously. (Galileo, *Opere*, 20:572)

After the signature, there follows an interesting post scriptum: "Would Your most Reverend Paternity advise as soon as possible whether the seal with the three fishes is the printer's, or is by Mr. Galileo, and would you see to writing to me adroitly what it means" (Galileo, *Opere*, 20:571–72). Even though Riccardi does not seem (or perhaps does not wish) to show any excessive worry, from his words it appears the pope had already shown a clear dissatisfaction with the content of the *Dialogue*. He had probably given a quick look here and there at the book, and he was not happy with it. And his unhappiness must have been even greater if such a reading had taken place because it was solicited by (or even with the assistance of) persons hostile to Galileo who were scandalized that permission had been given for the printing of a work so openly favorable to the Copernican system. Urban VIII would have above all become aware of how Galileo had introduced Urban's "theological" argument against the possibility of any proof of physical theories. He had seen it put in the mouth of Simplicius, the simpleton. No wonder the pope had found the *Dialogue* unsatisfactory! And perhaps somebody had cleverly insinuated the possibility that the dolphins that appeared in the frontispiece of the book could be an allusion to his nepotism. (In Italian, *delfino* [dolphin] is facetiously used to denote somebody who is promoted to a high position as a result not of personal merit but of protection by a superior authority. In the case of Urban VIII, the allusion was to his three nephews, two of whom— Antonio and Francesco Barberini—had been promoted to the position of cardinal, while the third one, Taddeo Barberini, had been named governor of Rome.)

Amidst the tension of that difficult moment of his pontificate, which had further exacerbated his natural sensitivity and self-esteem, Urban VIII could not but see in the *Dialogue* a betrayal of the trust he had placed in his friend and protégé of times past, a betrayal linked to that of Ciampoli, who now had added to his previous faults that of having kept silent about the fact that the *Dialogue* had been printed, as well as about its content.

Galileo, of course, knew nothing directly, about the letter of Riccardi to the Florentine inquisitor. The latter, however, must have "delicately" informed him that the sale of his book had been halted while waiting for it to be corrected. He received more details some weeks later from a long letter that Magalotti sent to Guiducci on August 7, and that the latter forwarded to Galileo (Galileo, *Opere*, 14:368–71). Magalotti mentioned rumors circulating in Rome of a possible correction, suspension, or even prohibition of the *Dialogue*. After mentioning the pope's reaction, he added:

> This is the pretext; but the real fact is that the Jesuit Fathers are working most valiantly in an underhanded way to get the work prohibited. The Reverend Father's [Riccardi's] own words were: "The Jesuits will persecute him most bitterly." (Galileo, *Opere,* 14:370; trans. Santillana, *The Crime of Galileo,* 190)

These words by Magalotti are generally quoted as proof of an action by the Jesuits that was supposedly the origin of the prohibition of the *Dialogue* and even of Galileo's condemnation. This matter will be discussed at the end of the present chapter. Here it seems to me necessary to stress the fact that Magalotti had grounded his judgment on Riccardi's affirmation. And the latter did not speak of a persecution already in existence but of one that he foresaw as certain in the future. From these words alone, therefore, one cannot conclude that the storm against Galileo had been aroused by the Jesuits. As for Scheiner, from the letter sent two days before (August 5) by Campanella to Galileo, it appears that he had not yet read the *Dialogue*. On the other hand, Magalotti himself, after the above mentioned words concerning the Jesuits, had added: "Besides, one cannot deny that His Holiness is of an absolutely contrary opinion." Without such a personal conviction of Urban VIII, the accusations made against the *Dialogue* would certainly not have had their desired effect.

Not satisfied at having stopped a further distribution of the *Dialogue*, Urban VIII had asked Riccardi to make sure that all the copies of it already in circulation be recovered. Almost at the same time, at the beginning of August, he had decided to entrust the examination of the *Dialogue* to "a commission of persons well versed in that profession, in the presence of Cardinal (Francesco) Barberini." This was, in fact, what Ambassador Niccolini reported to Cioli, the secretary of the grand duke, on August 15, adding that those persons were "not all well disposed towards Mr. Galileo" and that he intended to speak about the matter with the cardinal. Campanella too had heard this commission spoken of. He informed Galileo about it on August 21, affirming that among those "theologians angrily eager to prohibit the *Dialogue*," there were "Dominicans, Jesuits, Theatines and secular priests." He advised that pressure be exerted, through the grand duke, in order to have included in that commission Father Castelli and himself, in order to rebalance it as far as possible.

Niccolini, in fact, met Francesco Barberini a few days later. To his request to have included "uncommitted persons" in the commission for the examination of the *Dialogue,* the cardinal limited himself to the promise of reporting about it to the pope. He added, however, that it was a matter concerning a "friend of His Holiness, by whom he is loved and esteemed" and towards whom he himself had "good will" (Galileo, *Opere,* 14:375).

In a second, very long letter to Guiducci of September 4, Magalotti gave more accurate information on the charges against the *Dialogue,* again received from Riccardi. One of them was that the preface had been printed with characters different from those of the rest of the book, thus giving the impression of extraneous material. Another one, more serious, was the fact that Urban VIII's argument had been put in the mouth of Simplicius, "a personage very little esteemed in the whole treatment, in fact, rather treated with derision and mockery" (14:379). Magalotti was, however, rather optimistic and hoped that with some corrections and explanatory notes, the book would be put back in circulation. At any rate, Magalotti recommended great prudence and patience and counseled absolutely against trying to force the course of events with inopportune interventions, especially with Urban VIII.

On that same day, Niccolini had gone to Urban VIII in order to deal, most probably, with the question of a subject of the grand duke, one Mariano Alidosi. He had been accused to the Holy Office, but the grand duke had refused to have him judged, as requested, in Rome. Giving his report about that encounter, Niccolini wrote:

> While we were discussing those delicate subjects of the Holy Office,
> His Holiness exploded into great anger, and suddenly he told me
> that even our Galilei had dared entering where he should not have,
> into the most serious and dangerous subjects which could be stirred

up at this time. (Galileo, *Opere,* 14:383; trans. Finocchiaro, *Galileo Affair,* 229)

Niccolini attempted to plead Galileo's case, remarking that the latter had printed his book with the needed approval of the ministers of Urban VIII. But Niccolini had touched the wrong key:

> He [Urban VIII] answered, with the same outburst of rage, that he had been deceived by Galileo and Ciampoli, that in particular Ciampoli had dared to tell him that Mr. Galilei was ready to do all His Holiness ordered and that everything was fine, and that this was what he had been told, without having ever seen or read the work; he also complained about the Master of the Sacred Palace, though he said that the latter himself had been deceived by having his written endorsement of the book pulled out of his hands with beautiful words, by the book being then printed in Florence on the basis of other endorsements but without complying with the form given to the inquisitor, and by having his name printed in the book's list of imprimaturs even though he has no jurisdiction over publications in other cities. (Galileo, *Opere,* 14:384; trans. Finocchiaro, *Galileo Affair,* 229–30)

Even Niccolini's request that Galileo be granted the opportunity to justify himself before the commission of theologians that had to examine the *Dialogue,* was met with a flat denial by the pope. The ambassador insisted again, affirming he did not believe the pope would allow the condemnation of a book already approved, before hearing Galileo. Urban VIII answered that

> this was the least ill which could be done to him and that he should take care not to be summoned by the Holy Office; that he has appointed a Commission of theologians and other persons versed in various sciences, serious and of holy mind, who are weighing every minutia, word for word, since one is dealing with the most perverse subject one could ever come across; and again that his complaint was to have been deceived by Galileo and Ciampoli. (Galileo, *Opere,* 14:384; trans. Finocchiaro, *Galileo Affair,* 230)

Urban VIII further stressed that "every civility" had been employed with Galileo, since his case had not been entrusted to the Holy Office, as usually done, but to a specially created commission. This was a proof that one had used "better manners with Galileo, than the latter had used with His Holiness, who had been deceived" (Galileo, *Opere,* 14:384; trans. Finocchiaro, *Galileo Affair,* 230–31).

For Niccolini the experience had been a bitter one. He therefore preferred not to intervene again with Urban VIII, in spite of new instructions to do so that had come from Florence. Even Father Riccardi, whom he had consulted, counseled in the most absolute fashion against any other interventions with the pope, "so as not to end up by ruining poor Mr. Galilei and by breaking off with His Holiness" (Galileo, *Opere,* 14:388). In his report to Cioli of September 11, Niccolini added that Riccardi was reviewing the work and intended to bring it to the pope, assuring him that it could be "allowed to circulate and that His Holiness now had the opportunity of using his customary mercy with Mr. Galileo" (14:389).

Riccardi sought also to reassure Ambassador Niccolini with respect to the members of the commission to which the examination of the *Dialogue* had been entrusted. One of them was he himself, Riccardi, who had every reason to defend Galileo, the more so since he had "endorsed the book." Another member was the pope's theologian (Agostino Oreggi), who "truly has good will." A third, proposed by Riccardi himself, was in all probability the Jesuit Melchior Inchofer, who according to Riccardi was "his [Riccardi's] confidant" and was moved by "correct intentions." On this point, Riccardi had been mistaken (or was not sincere). As a matter of fact, Inchofer was a convinced anti-Copernican, as shown by his *Tractatus Syllepticus,* published the following year. And his report on the *Dialogue,* at the time of the second commission in April 1633, would show how much he was ill-disposed towards Galileo. Was there also a fourth member of the commission, whom Riccardi had not mentioned? Campanella had spoken also of "Theatines." It is thus possible that the Theatine Zaccaria Pasqualigo, who will take part, together with the others already mentioned, in the second commission, had also been included in the first one.

As for Campanella's proposal to include himself and Castelli in the commission, Riccardi affirmed that this was not possible, even more so given the suspicion with which both of them were seen in Rome. Castelli was no doubt suspected because of his well-known friendly relations with Galileo, and perhaps also for having given a reassuring report on the *Dialogue* to Cardinal Francesco Barberini from whom, as we know, he had borrowed a copy of the book. Concerning Campanella, his *Apologia pro Galileo,* published in Germany in 1622, had been prohibited that same year by the Roman authorities. And because of this work he had been (wrongly) considered as a supporter of the Copernican ideas of Galileo.

Together with this more or less "reassuring" news (at least as intended by Riccardi) he had added in all secrecy another piece of news of a quite different tenor. He had in fact confided to Niccolini that

> in the files of the Holy Office they have found something which alone is sufficient to ruin Mr. Galilei completely; that is, about twelve years ago, when it became known that he held this opinion and was sowing it in Florence, and when on account of this he was called to Rome, he was prohibited from holding this opinion by the Lord Cardinal Bellarmine, in the name of the Pope and the Holy Office. (Galileo, *Opere,* 14:389; trans. Finocchiaro, *Galileo Affair,* 233)

Riccardi had not reported exactly either the circumstances or the content of the document recovered from the archives of the Holy Office. As is clear from the proceedings of Galileo's trial, it was no doubt the document of February 26, 1616, with the precept by Bellarmine, followed by the Segizzi injunction. The inexactitude of Riccardi's mention of it would appear to be a sign of the fact that the recovery of the document was not made by the commission. Otherwise Riccardi, as a member of it, would have been better informed. That discovery was perhaps made by some member of the Holy Office, hostile to Galileo and concerned and dissatisfied that the question of the *Dialogue* could be resolved outside the Holy Office. And some archivist, well-acquainted with the existence of the Galileo dossier, could have easily helped him.

Whatever the case, the precept of Bellarmine alone, mentioned by Riccardi, was sufficient to prove Galileo culpable, in the case the commission concluded that in the *Dialogue* he had defended the Copernican theory. There was the further aggravating circumstance that

Galileo, at the moment of the request for the *imprimatur* for the *Dialogue,* had kept silent with Riccardi about the existence of the precept. Such an *imprimatur* thus had been fraudulently obtained. It was indeed in this sense that Riccardi judged the discovery as able to completely ruin Galileo.

That Galileo had in fact defended the Copernican theory was the conclusion of the commission. It judged, accordingly, that the question of the *Dialogue* should have been entrusted to the Holy Office, so that a decision could be taken by it on the matter. This was what Urban VIII made known to Niccolini by one of his secretaries and what he confirmed directly to the ambassador three days later. By now the machinery of the Holy Office had started rolling, and it would not have been easy for anyone to stop it, not even Urban VIII. On September 23 there was a meeting of the Congregation of the Holy Office, with the pope and eight cardinals present. During the meeting, a report was read on the facts concerning the printing of the *Dialogue,* at the end of which, in a synthetic form, were contained the responses of the commission that had examined it. And there was joined to it that which will become a further, serious charge, namely the mention of the Segizzi injunction:

> The author had the command laid upon him by the Holy Office in the year 1616 to relinquish altogether the said opinion that the Sun is the center of the world and that the Earth moves and henceforth he is not to hold, teach, or defend it in any way whatsoever, verbally or in writing, otherwise proceedings will be taken against him by the Holy Office. To which injunction he acquiesced and promised to obey. (Galileo, *Opere,* 19:279)

As one can see, in summary form the document of February 26, 1616, is quoted here, with the injunction by Segizzi, but without naming him. It is possible that the contradiction of Segizzi's action with the instructions he had received from Cardinal Millini and with the report given to the Holy Office by Bellarmine about the precept having been carried out, had not been missed by those who had recovered the document and perhaps by some of the cardinal inquisitors. Even the summary of the Galileo trial, submitted at the end of it for the final decision by the pope, will not mention Segizzi, attributing the whole injunction to Bellarmine, as Riccardi had already done when speaking with Magalotti.

Apart from the question of its author, the existence of the document, by now known to Urban VIII himself, must have provoked a further and even deeper resentment in him. The pope must have felt once more that he had been deceived by his old "friend," who in the long conversations that had taken place with him had not made any mention of a "precept" received from Bellarmine and even less of one in a more rigorous form. Thus, Urban VIII's decision could not be but the most severe one. Galileo had to be called to Rome in order to be subjected to a personal procedure by the Holy Office. That was, after all, what was foreseen in the instructions of Cardinal Millini as well as in the injunction by Segizzi.

In fact, at the conclusion of the meeting, Urban VIII

> ordered that a letter be sent to the inquisitor of Florence so that he would tell the same Galileo, in the name of the Sacred Congregation, to appear within the month of the coming October in Rome, before the Commissary General of the Holy Office, and that he receive from him the promise to obey that command, the promise to be made in the presence of witnesses who, if necessary, will be able to testify in case he does not wish to accept the command and does not promise to obey. (Galileo, *Opere,* 19:279–80)

On October 1 Galileo was, in fact, called by the Florentine inquisitor, who delivered the precepts to him according to the instructions he had received. Galileo declared that he willingly accepted them. In reality, this order took him completely by surprise and deeply disturbed him. In order to avoid a formal trial in Rome, he sought help from the grand duke and from Cardinal Francesco Barberini. He wrote a long letter to the latter, in which he proposed to prepare "for the Fathers of the Holy Office" an account about all he had written and done concerning the Copernican opinion, trusting that this report would be adequate to persuade them of his innocence. As an alternative, he asked to be allowed to justify himself before the inquisitor, the nuncius, the archbishop, and other members of the Florentine clergy. And he concluded:

When neither my advanced age nor my many bodily ills, nor my troubled mind, nor the length of a journey made most painful by the current suspicions are judged by this sacred and high Court to be sufficient excuses for seeking their dispensation or postponement, I will take up the journey, preferring obedience to life itself. (Galileo, *Opere,* 19:410)

Galileo's friends had also become involved in helping him. Castelli, in particular, had taken steps with Father Riccardi and the commissary of the Holy Office, Ippolito Maria Lanci. He himself informed Galileo about this with his letter of October 2. According to Castelli, the commissary had frankly admitted that the question of the Earth's movement "should not be concluded with the authority of the Sacred Letters" and had even declared "that he wanted to write about it, and that he would have made it known to me." These words seem to be an indication of the disparity of views that existed even within the Holy Office concerning the Copernican question. Lanci, however, will resign his position towards the end of the year, thus depriving Galileo of the chance to have, at the moment of the trial, an investigating judge (that was in fact the commissary's task) sympathetic to him.

For his part, Niccolini, under instructions from the grand duke, had busied himself to bring it about that Galileo should not have to go to Rome. After consultation with Castelli, he delivered to Cardinal Francesco Barberini the letter Galileo had written to him, and on November 13 he tried to plead Galileo's cause with Urban VIII himself. The pope however was unmoved, affirming that "it was necessary to examine him [Galileo] personally, and that God would hopefully forgive him his error of having gotten involved in an intrigue like this, after His Holiness himself (when he was cardinal) had delivered him from it." Niccolini had again insisted on the fact that the *Dialogue* had been approved, but, as he reported to Cioli:

I was interrupted by being told that Ciampoli and the Master of the Sacred Palace had behaved badly and that subordinates who do not do what their masters want are the worst possible servants; for, when asking Ciampoli many times what was happening with Galilei, His Holiness had never been told anything but good and had never been given the news that the book was being printed, even when he was beginning to smell something. Finally, he reiterated that one is dealing with a very bad doctrine. (Galileo, *Opere,* 14:428–429; trans. Finocchiaro, *Galileo Affair,* 239)

Urban VIII thus showed himself more than ever convinced that he had been tricked by his most trustworthy aides in the entire question of the *Dialogue,* especially by Ciampoli. That conviction had probably been the death blow with respect to the latter, who had already fallen into disgrace as of March for other reasons, as we know. In fact, he had to leave Rome on November 23, to go as governor to a small town of the Marches, Montalto, under orders not to come back to Rome. From Montalto he would later go, always as governor, to Nocera Umbra, Fabriano, and finally to Jesi, where he would die in 1643.

The unrelenting attitude shown by Urban VIII to Niccolini was the consequence of a decision that he had personally taken during the meeting of the Holy Office two days before. During the course of that meeting consideration had been given to the request of the grand duke of Tuscany that Galileo, because of his advanced age, be permitted not to come to Rome. "His Most Holy Person," report the minutes of the meeting, "has not wished to make any concession, but he ordered me to write that he must obey and [to write] to the Inquisitor that he would force him to come to Rome" (Galileo, *Opere,* 19:280).

The Florentine inquisitor carried out the assigned task but granted to Galileo a one month's delay, in the presence of a notary and two witnesses. Galileo tried again to procrastinate. On November 17 he sent to Rome a declaration of his bad health, prepared by three doctors, which was read during the meeting of the Holy Office of November 30. By now, Urban VIII had very definitely lost his patience. That is very obvious in the formulation of the decision taken in this regard, which was extremely severe. In order to verify if, in fact, Galileo could not make the trip to Rome without danger to his life, the commissary of the Holy Office would have to be sent, with doctors, at Galileo's expense. If they concluded that Galileo was in such a state as to be able to start the trip, they should send him imprisoned and in chains. If on the other hand it was necessary to put off his departure, once he recovered he had to be transported to Rome equally in prison and in chains.

Urban VIII had probably wanted to make such a display of severity in order to show to the cardinals of the Holy Office that he was still firmly in command. In fact, the instructions sent by the brother of the pope, Cardinal Antonio Barberini, to the Florentine inquisitor, were more moderate. The inquisitor had to tell Galileo that if he did not obey right away, the commissary with doctors would be sent to Florence and Galileo would be transported to Rome imprisoned and in chains.

Inquisitor Egidi read the injunction to Galileo, who affirmed that he was ready to obey, even though he denied that his illness was a subterfuge. After the meeting with Egidi, Galileo tried a last intervention with the grand duke. The response he obtained, through Secretary Cioli, was full of sympathy but clear. Galileo had finally to "obey the higher tribunals." But to sweeten the bitter pill, the grand duke put one of his litters at Galileo's disposal and arranged that he would be hosted in Rome in the residence of Ambassador Niccolini. No doubt, the grand duke wanted thus to show also that Galileo continued to enjoy his esteem and trust.

Galileo realized that there was nothing left to do but to go, and on January 20 he finally set out on his trip to Rome. Because the plague was still in existence, he had to spend a quarantine period in Ponte a Cecina, on the border between the Grand Duchy of Tuscany and the Papal States. The stop was longer than foreseen, in an uncomfortable lodging and with "wine, bread, and eggs" as the only food. It was only on February 13 that he reached Rome, taking up lodging at the Tuscan embassy. After all the anxiety and the physical discomfort that had tormented him during the past weeks, there was now at least, to give him again hope, all the warmth of the friendship of Ambassador Niccolini and of his wife Caterina Riccardi. And indubitably he was also comforted by the fact that the pope had allowed him to stay at the embassy instead of in the prisons of the Holy Office.

The day after his arrival in Rome, Galileo paid a visit to Monsignor Boccabella, who had just left the office of assessor of the Holy Office. Obviously, he wanted to seek advice and possibly obtain some information about what was ongoing. Upon Boccabella's advice, Galileo went to visit the new assessor, Pietro Paolo Febei, and then to the new commissary of the Holy Office, Vincenzo Maculano, whom he unfortunately was unable to find.

Cardinal Francesco Barberini, however, let him know a few days later that it was better that Galileo did not pay or receive visits and stayed instead at "home retired, waiting to be told something." Obviously, the Holy Office wanted him to know that even though not in seclusion, he was not free to move and act as he wanted. The same advice was made known to him by Commissary Maculano, who stressed that it was not a question of an order but of "friendly advice." Galileo understood right away what this "friendly advice" meant and scrupulously observed it.

Even though the Holy Office did not provide any information to Galileo, he continued little by little to regain courage and hope. The advice he had received from Cardinal Francesco Barberini and Maculano had been formulated with kindness. And the same benignity had been shown to him by the consultor of the Holy Office, Lodovico Serristori, during the two visits he made to Galileo. These visits, probably mandated during the meeting of the Holy Office of February 16, were aimed at sounding out Galileo, in order to determine more concretely the way to deal with him. Galileo's resurgence of optimism is apparent in his words to Cioli, in the letter of February 19: "this seems to be a beginning of procedure which is very gentle and kind, and completely unlike the threatened ropes, chains, and prisons, etc." (Galileo, *Opere,* 15:44).

For his part, Niccolini had gone to visit Cardinals Bentivoglio and Scaglia, members of the Holy Office, and had found them very well disposed towards Galileo. Cardinal Guido Bentivoglio had followed Galileo's lectures at Padua and, in fact, he would try to help him during the trial, as he himself affirmed in his memoirs. Desiderio Scaglia was a Dominican. From Castelli we know that he had started reading the *Dialogue* with his help, "forming an opinion of it, if not altogether contrary, at least very different and far from the one he had previously held."

On February 26 Ambassador Niccolini spoke with Urban VIII himself. As he reported the day after to Cioli, Urban VIII had underlined the fact that, in being allowed to live in the Tuscan embassy, Galileo had been accorded an altogether exceptional treatment. And this had been done only out of respect for the grand duke. After having informed Niccolini that the trial was still in the process of preparation and that he did not know if it could be brought to a quick conclusion, the pope had added that, in short, Mr. Galilei had been ill-advised to publish these opinions of his, and it was the sort of thing for which Ciampoli was responsible; for although he claims to want to discuss the earth's motion hypothetically, nevertheless when he presents the arguments for it he mentions and discusses it assertively and conclusively; furthermore, he had also violated the order given him in 1616 by Lord Cardinal Bellarmine in the name of the Congregation of the Index [*sic*]. (Galileo, *Opere*, 15:56; trans. Finocchiaro, *Galileo Affair*, 245)

After some time, Galileo had still received no communication from the Holy Office. He sought to console himself thinking that this delay might have contributed to bringing about the nullification of so many charges against him. But he was wrong. Niccolini, as a consummate diplomat, had written to Cioli that the Holy Office could not allow the trial to fade into nothing, even if Galileo would be able to bring forth in his defense proofs that were satisfactory. And he obtained a confirmation of his conviction on the occasion of a new meeting with Urban VIII, on March 13. The pope made it known to him that he did not think it was possible to avoid having Galileo called to the Holy Office for trial. He promised, however, that "he would see to it that certain rooms, which are the best and the most comfortable in the palace, were assigned to him" (Galileo, *Opere*, 15:67–68).

The two months of uncertainty finally came to an end. On April 6, Niccolini was summoned by Cardinal Francesco Barberini, who informed him that by order of the pope and of the Congregation of the Holy Office Galileo should be summoned to the same Holy Office. He, however, confirmed what Urban VIII had promised, namely, that Galileo "would be kept there not as if in prison nor in secret, as was usually done with others, but that he would be provided with good rooms that would perhaps even be let open" (15:85).

Three days later, Niccolini went to thank the pope for the special respect shown towards the grand duke, but he found him more than ever fixed in the position taken: "His Holiness complained that he [Galileo] had entered into that matter that for him [the pope] is still a most serious matter and one that has great consequences for religion" (15:85). Worried about such rigidity on the part of Urban VIII, in informing

Galileo of his imminent summons to the Holy Office, he recommended that he not try to defend his positions, but "that he submit to what he might see they would want him to believe and hold in that particular about the mobility of the Earth." And in his report to Cioli, Niccolini added: "He is extremely afflicted by this; and judging by how much I have seen him to go down since yesterday, I have very serious worries about his life" (15:85).

On April 12 Galileo was finally summoned to the Holy Office, as Niccolini wrote to Cioli four days later. The commissary had received him "in a friendly manner and had him lodged in the chambers of the prosecutor of that tribunal" so that he not only resides "among the officials, but he is free to go out into the courtyard of that house" (15:94). Niccolini added that they allowed his domestic to serve him and stay with him and that the servants of the embassy would bring him his food in the morning and in the evening.

On the same day the first interrogation took place, known to us through the proceedings of Galileo's trial. Besides Maculano, who acted as investigatory judge, were present his assistant, the prosecutor of the Holy Office, Carlo Sinceri, and a secretary, who transcribed the minutes. The interrogation started with a series of questions about the events of 1616 and about the decisions then taken with regard to the Copernican issue. They were intended as a preparation for the crucial question, which was aimed at obtaining from Galileo a confirmation of the content of the document recovered six months earlier. Commissary Maculano formulated it as follows: "That he declare what had been decided and communicated to him at that time, namely in the month of February 1616." Galileo answered:

> In the month of February 1616, Lord Cardinal Bellarmine told me that since Copernicus's opinion, taken absolutely, was contrary to Holy Scripture, it could neither be held nor defended, but it could be taken and used suppositionally. In conformity with this I keep a certificate by Lord Cardinal Bellarmine himself, dated 26 May 1616, in which he says that Copernicus's opinion cannot be held or defended, being against Holy Scripture. I present a copy of this certificate, and here it is. (Galileo, *Opere*, 19:339; trans. Finocchiaro, *Galileo Affair*, 259)

The difference between the content of the document signed by Bellarmine, presently before his eyes, and the one newly found from the archives could not escape the commissary. The latter document, as we know, spoke of the intervention of Commissary Segizzi right after the admonition given by Bellarmine. Nothing on such an intervention was mentioned in Bellarmine's certificate. This is why Maculano further asked: "Whether, when he was notified of the above mentioned matters, there were others present, and who they were." Galileo admitted that "there were present some Dominican Fathers." But he added that he did not know them and had never seen them since. Maculano pressed: "Whether at that time, in the presence of those Fathers, he was given any injunction concerning the same matter, and if so what." Galileo answered:

> As I remember it, the affair took place in the following manner. One morning Lord Cardinal Bellarmine sent for me, and he told me a certain detail that I should like to speak to the ear of His Holiness before telling others; but then at the end he told me that Copernicus's opinion could not be held or defended, being contradictory to Holy Scripture. I do not recall whether those Dominican Fathers were there at first or came afterwards; nor do I recall whether they were present when the Lord Cardinal told me that the said opinion could not be held. Finally, it may be that I was given an injunction not to hold or defend the said opinion, but I do not recall it since this is something of many years ago. (Galileo, *Opere*, 19:339–40; trans. Finocchiaro, *Galileo Affair*, 259).

We do not know what the "detail" alluded to here by Galileo could have possibly been. It was certainly not the intervention of Cardinal Maffeo Barberini, which took place four days later on the occasion of the meeting of the Congregation of the Index of March 1, 1616, in order to avoid the condemnation of the Copernican theory as heretical. It is however possible that the same cardinal had already intervened against such a hasty condemnation or perhaps in favor of Galileo himself on the occasion of the consistory on February 24, or speaking privately with Bellarmine around the same time. At any rate, Bellarmine could have hinted at this "detail" at the beginning of Galileo's summons, in order to make less painful what he was going to tell him. As if to recall to Galileo's memory the particulars of his summoning by Bellarmine, Maculano said that the injunction given to him in the presence of witnesses contained the words: "that he cannot in any way whatsoever hold, defend, or teach that opinion" and he asked Galileo whether he remembered it now and from whom he had been so ordered. Galileo's reply corresponds to what he had already declared:

> I do not recall that such injunction was given me any other way than orally by Lord Cardinal Bellarmine. I do remember that the injunction was that I could not hold or defend, or maybe even that I could *not teach.* I do not recall, further, that there was the phrase *in any way whatever,* but maybe there was; in fact, I did not think about it or keep it in mind, having received a few months thereafter Lord Cardinal Bellarmine's certificate dated 26 May which I have presented and in which is explained the order given to me not to hold or defend the said opinion. Regarding the other two phrases in the said injunction now mentioned, namely *not to teach* and *in any way whatever,* I did not retain them in my memory, I think because they are not contained in the said certificate, which I relied upon and kept as a reminder. (Galileo, *Opere,* 19:340; trans. Finocchiaro, *Galileo Affair,* 260)

Galileo had made here, maybe inadvertently, a fatal mistake. He had now admitted having received from Cardinal Bellarmine an *injunction* not to hold or defend the Copernican opinion. Such admission neutralized for the most part the advantage he had had up to now in producing his certificate. An injunction was much more than the simple *information* that resulted from that document. Galileo, moreover, had admitted at least the possibility of having received—always by Bellarmine—the order *not to teach* and *in any way whatever* the Coperican doctrine, and these words too were not included in that certificate. We will see later on how this admission by Galileo would be used in the "Summary of the Trial" for the neutralization of the Bellarmine document.

Even Maculano seems to have been aware of it, and therefore satisfied with such a confession. This admission was enough for Maculano, even leaving aside the document of the Holy Office archives, which was even more precise and stronger. And so he insisted: "Whether, after the issuing of the said injunction, he received a special permission to write the book identified by himself, which he later sent to the printer." There stood the crux of the question and Galileo knew it. But by now he had decided upon his line of defense. And so he responded:

> After the above-mentioned injunction I did not seek permission to write the above-mentioned book that I have identified, because I do not think that by writing this book I was contradicting at all the injunction given me not to hold, defend, or teach the said opinion, but rather that I was refuting it. (Galileo, *Opere*, 19:340)

To refute it! This statement of Galileo must have caused the commissary to start, since he knew well what the commission to whom the examination of the *Dialogue* had been entrusted had thought of this "refutation." Maculano, however, preferred not to start arguing with Galileo on this point, knowing that a new commission would again examine the *Dialogue*. He thus proceeded to the second question, the one of the permission for the printing of the *Dialogue*. And so he asked: "Whether when he asked the Master of the Sacred Palace for permission to print the above-mentioned book, he revealed to the same Most Reverend Father Master the injunction previously given to him concerning the directive of the Holy Congregation, mentioned above." Galileo's reply was in line with the previous one:

> When I asked him for permission to print the book, I did not say anything to the Father Master of the Sacred Palace about the above-mentioned injunction because I did not judge it necessary to tell it to him, having no scruples since with the said book I had neither held nor defended the opinion of the earth's motion and sun's stability; on the contrary, in the said book I show the contrary of Copernicus's opinion and show that Copernicus's reasons are invalid and inconclusive. (Galileo, *Opere*, 19:341; trans. Finocchiaro, *Galileo Affair*, 261–62)

With these declarations, Galileo had obviously hoped to save himself and the *Dialogue* based upon his protests on the nonconclusiveness of the arguments in favor of the Copernican theory, here and there repeated in the book, and especially in the Urban VIII argument, quoted by him at the end of the *Dialogue*. Once again, Maculano did not want to start arguing with him. He knew the different opinion expressed on the matter by the special commission more than six months before, as well as the fact that a new commission would now examine the *Dialogue* even more exhaustively. Galileo signed the minutes and he took the oath to observe secrecy. He was then told to stop in at the Holy Office, at the prosecutor's apartment. Galileo had undoubtedly hoped to be able to return to Palazzo Firenze. But the courteous attitude of the commissary and the special respect shown to him must have made his disillusionment less bitter. And perhaps he had found some consolation in the fact that it would only be a matter of a few days.

In fact, on April 23 he was still at the Holy Office. In a letter of that day to Geri Bocchineri, private secretary of the grand duke, Galileo related that the commissary and the prosecutor had come to pay a visit to him, while he was in bed with a leg pain, and they had assured him that they had the intention to complete the interrogations as soon as he had recovered, insisting several times that he should be of happy and tranquil mind. This gave to Galileo new hope that his innocence and sincerity would come to be known.

Even Niccolini seemed fundamentally optimistic and was convinced that Galileo would be let "free" as soon as the pope would return from his villa in Castel Gandolfo for the feast of the Ascension, which that year fell on May 5. In reality, the situation was very different. Since Galileo had stated that he had not wished to defend the Copernican doctrine in the *Dialogue,* the Holy Office had asked the same theologians who had examined it six months before to answer a precise question. Had Galileo held, taught, and defended the opinion of the mobility of the Earth and of the immobility of the Sun? Evidently, they wanted to make it clear once and for all whether Galileo had violated the inquisitorial precept imposed upon him in the more severe form by Segizzi. If proven, Galileo would also be guilty of a violation of the decree of the Congregation of the Index of 1616. On April 17 the theologians gave their answer, stating that Galileo, by writing the *Dialogue,* had in effect transgressed the command of the Holy Office.

More precisely, he had certainly taught and defended the opinion of the motion of the Earth and of the immobility of the Sun. And he was also very suspect (Pasqualigo) and even "vehemently suspect" (Inchofer) of having held it. According to the practice of the Inquisition of these times, the technical term "vehemently suspect" implied that the theory of Copernicus, if not formally a heresy (Inchofer certainly held this), was at least an error in faith, which was sufficient to make the "vehemently suspect" into a "vehemently suspect of heresy." Such suspicion could not be canceled except by an abjuration, and it carried with it imprisonment.

The response of the three theologians was examined and accepted by the Holy Office in the meeting of April 21, which is known from a letter recently found by Ugo Baldini in the archives of the former Holy Office. It was sent the day after by Maculano to Cardinal Francesco Barberini, who was at that time at Castel Gandolfo with Urban VIII. In the central passage of the letter, which concerns Galileo, Maculano wrote:

> Last night Signor Galileo was torn by pains that assailed him and was groaning also this morning. Such is his state that, having visited him twice, he received the greatest solace upon hearing that we wanted his case to be handled as soon as possible. I truly think that this should be done considering the advanced age of this man. Already yesterday the Congregation considered the book and it was determined that in it the opinion disapproved and condemned by the Church is defended and taught, and that, moreover, the author makes himself suspect of personally holding that opinion. Since that is the way the question stands, it could quickly be treated as a case to be handled expeditiously, for which I await Your Excellency's directive so that I can carry it out punctually. (Beretta, "Un nuovo documento," 640)

As we see, Maculano shows that he is personally interested in a speedy conclusion of the trial of Galileo, at least in part from an apparently sincere regard for Galileo's "advanced age." And he considers that a decisive step in this direction has been taken with the conclusion by the Holy Office as to the content of the *Dialogue* and the suspicion, by now quite clear, that the author holds the opinion of Copernicus, "already disapproved and condemned by the Church." It must be noted, however, that Maculano does not prefix the qualification "vehemently" to the word "suspect." Since that qualification of "vehemently suspect" had been given only by Inchofer, it is possible that in the meeting of the Holy Office, mentioned by Maculano, that qualification did not gain the unconditional approval of the cardinals. Or maybe Maculano, while making his report to Francesco Barberini, had wanted to circumvent the seriousness of the conclusion, hoping that a solution of the Galileo case, not only quick, but also more benign, was still possible, avoiding abjuration and prison. It is possible that he knew that the "cardinal nephew" was also inclined to such a solution. Of particular importance is the final statement of Maculano where he says that "he awaits a directive" from the cardinal, so that he might "carry it out punctually."

This document, therefore, provides evidence of a close rapport between Maculano and Cardinal Francesco Barberini, and that the latter was disposed to intervene personally in favor of a rapid conclusion to Galileo's case. That Francesco Barberini, in fact, sent the directives requested by Maculano seems clear from the final words of the following letter sent by Maculano to the cardinal on April 28, already published by Favaro:

> Yesterday, in accordance with the orders of His Holiness, I reported on Galileo's case to the Most Eminent Lords of the Holy Congregation by briefly relating its current state. Their Lordships approved what has been done so far, and then they considered various difficulties in regard to the manner of continuing the case and leading it to a conclusion; for in his deposition Galileo denied what can be clearly seen in the book he wrote, so that if he were to continue in his negative stance it would become necessary to use greater rigor in the administration of justice and less regard for all the ramifications of this business. Finally I proposed a plan, namely that the Holy Congregation grant me the authority to deal extrajudicially with Galileo, in order to make him understand his error and, once having recognized it, to bring him to confess it. The proposal seemed at first too bold, and there did not seem to be much hope

of accomplishing this goal while there existed the road of convincing him with "reasons"; however, after I mentioned the basis on which I proposed this, they gave me the authority. In order not to lose time, yesterday afternoon I had a discussion with Galileo, and, after exchanging innumerable arguments and answers, by the grace of the Lord I accomplished my purpose: I made him grasp his error, so that he clearly recognized that he had erred and gone too far in his book; he expressed everything with heartfelt words, as if he were relieved by the knowledge of his error; and he was ready for a judicial confession. However, he asked me for a little time to think about the way to render his confession honest, for in regard to the substance he will hopefully proceed as mentioned above.

I have not communicated this to anyone else, but I felt obliged to inform Your Eminence immediately, for I hope His Holiness and Your Eminence will be satisfied that in this manner the case is brought to such a point that it may be settled without difficulty. The Tribunal will maintain its reputation; the culprit can be treated with benignity; and, whatever the final outcome, he will know the favor done to him, with all the consequent satisfaction one wants in this. I am thinking of examining him today to obtain the said confession; after obtaining it, as I hope, the only thing left for me will be to question him about his intention and allow him to present a defense. With this done, he could be granted imprisonment in his own house, as Your Eminence mentioned. And to you I now express my humblest reverence. (Galileo, *Opere,* 15:106–7; trans. Finocchiaro, *Galileo Affair,* 276–77, with slight modifications)

How are we to interpret the contents of this letter? It seems to be consistent with the preceding letter in its attempt to overcome the impasse into which Galileo's trial had fallen. On the one hand, as the result of the response of the theologians, Galileo had already appeared insincere in his claim that he had not wanted to defend or to teach Copernicanism in the *Dialogue* and that he did not personally hold it. And that would have justified a "greater rigor in rendering justice," which in the practice of the Inquisition of the time meant a rigorous examination "on his intention," which was usually accompanied by torture. But given

Galileo's advanced age, torture would not have gone beyond the threat of it. And Galileo was as always the "first Philosopher and Mathematician" of the grand duke of Tuscany and was being solicitously helped by him through Ambassador Niccolini. Furthermore, he was one of the most famous and respected men in the Europe of his day. All of this made up those considerations to which the Holy Office had thus far given its attention. Despite everything, it was an attention that could not be easily dismissed.

What was Maculano trying to accomplish with his proposal? From his own words it is clear that he intended first of all to get a confession from Galileo that would have allowed for a quick conclusion of the trial, by overcoming the obstacles to which we have already referred. And it was a question of a "benign" procedure, seeing that the confession would not have been extorted by the threat of torture but obtained through persuasion in the course of an extrajudicial conversation. The juridical confession, to be requested afterwards, would have thus been little more than a formality.

Was there an additional element in the leniency of Maculano's proposal? From the words "the culprit can be treated with benignity," as well as from what, as we shall see, Maculano will lead Ambassador Niccolini to understand later on, it appears that he was hoping that Galileo's confession would have made possible a solution of the trial less severe, making it even possible to avoid a public condemnation. Everything, of course, would depend on the theological censure that would be decided with respect to the Copernican theory. If that of "rash" (*temeraria*) were accepted, it would be possible to avoid an abjuration and imprisonment for Galileo. It is possible that this censure was the one that Maculano held personally, as it was for Urban VIII himself, at least in the past. But the commissary himself recognized in his letter that the final decision in the case belonged to the Holy Office.

Was such an initiative taken by Maculano on his own? The "foundation" of which he made mention and that was enough to overcome the doubts expressed on the matter by the cardinals of the Holy Office, must have carried significant weight. The support of a "cardinal nephew," such as Francesco Barberini, was certainly not indifferent. That weight, however, derived from the fact that he was seen by the other cardinals of the Holy Office, just because of his kinship with the pope, as the spokesman of the pope's desires if not of his will. And it is very probable that Urban VIII had made known to Maculano, directly through the cardinal nephew, his desire for a quick conclusion of the trial, in line with the many requests made on the matter by Ambassador Niccolini. Thus the commissary would have felt that in making the proposal his back was covered.

The end of the letter makes clear the real aim of the whole extrajudicial initiative. After the confession, states Maculano, "the only thing left for me will be to question him about his intention and allow him to present a defense." That agrees with the contemporary practice of the Inquisition, whereby the guilty party, having confessed, was also obliged to express his intentions, that is, the reasons why he had committed his error. Even the granting of a written defense was part of the procedure of the Inquisition. But what does the final statement mean, the one, that is, that "With this done, he could be granted imprisonment in his own house, as Your Eminence mentioned"? It seems that Maculano was alluding here to the conclusion of the trial in a benign form with no abjuration but only with house arrest and the condemnation of the *Dialogue,* which was inevitable.

In fact, the second interrogation of the trial, which Maculano hoped would finally have brought Galileo to the point of making a full confession, took place not on the same day, April 28, but two days later. Galileo stated that he had during those days reflected continuously, especially with respect to the injunction given to him in 1616 by order of the Holy Office "not to hold, defend, or teach in any way the opinion, by then condemned, of the mobility of the Earth and the stability of the Sun." And having reread the *Dialogue,* which he had not looked at for three years, so that he might examine whether he really failed to follow the command, he confessed that his writing

> appeared to me in several places to be written in such a way that a reader, not aware of my intention, would have had the reason to form the opinion that the arguments for the false side, which I intended to confute, were so stated as to be capable of convincing because of their strength, rather than being easy to answer. In par-

ticular, two arguments, one based on sunspots and the other on the tides, are presented favorably to the reader as being strong and powerful, more than would seem proper for someone who deemed them to be inconclusive and wanted to confute them, as indeed I inwardly and truly did and do hold them to be inconclusive and refutable. As an excuse for myself, within myself, for having fallen into an error so foreign to my intention, I was not completely satisfied with saying that when one presents arguments for the opposite side with the intention of confuting them, they must be explained in the fairest way and not be made out of straw to the disadvantage of the opponent, especially when one is writing in dialogue form. Being dissatisfied with that excuse, as I said, I resorted to that of the natural gratification everyone feels for his own subtleties and for showing himself to be cleverer than the average man, by finding ingenious and apparent considerations of probability even in favor of false propositions. . . . My error then, and I confess it, one of vain ambition, pure ignorance, and inadvertence. (Galileo, *Opere,* 19:342–43; trans. Finocchiaro, *Galileo Affair,* 278)

After having released this declaration, Galileo returned to his rooms and said:

And for greater confirmation that I neither did hold nor do hold as true the condemned opinion of the earth's motion and sun's stability, if, as I desire, I am granted the possibility and the time to prove it more clearly, I am ready to do so. The occasion for it is readily available since in the book already published the speakers agree that after a certain time they should meet again to discuss various physical problems other than the subject already dealt with. Hence, with this pretext to add one or two other Days, I promise to reconsider the arguments already presented in favor of the said false and condemned opinion and to confute them in the most effective way that the blessed God will enable me. So I beg this Holy Tribunal to cooperate with me in this good resolution, by granting me the permission to put it into practice. (Galileo, *Opere,* 19:344; trans. Finocchiaro, *Galileo Affair,* 278–79)

This deposition of his and especially the final addition have been the object of criticism, at times very severe, with respect to Galileo. Undoubtedly these statements of his were influenced by the state of tension of the last months and especially of the last weeks. But it seems to me exaggerated to consider them to be the fruit of a state of panic, in the face of an imprecise, pressing danger. After all, the conversation with Maculano and the deposition had taken place in a friendly atmosphere and therefore without special tension. In reality, with that proposal, for us absurd not to say degrading, to write against Copernicanism, he had perhaps deluded himself that he would be able to salvage the *Dialogue,* avoiding its unconditioned prohibition. The book could have been only suspended until it had been corrected, with the elimination of those parts that, through the excessive "subtlety and vanity" of the author, had given the impression that they constituted a confirmation of the Copernican theory. And the final doubts in this matter would have fallen away with the appendage of one or two additional days, which would have made absolutely clear that the intention of the *Dialogue* was not only non-Copernican but, as a matter of fact, anti-Copernican. What would be the thoughts of those who one day might read the *Dialogue* so emasculated and supplied with such an unbelievable addition? Galileo probably trusted that at least the more intelligent among them would have been able to decipher the "mystery."

Was the declaration that Galileo made (apart from his later proposal) in line with what Maculano expected? The answer would appear to be yes. It is probable that also in the course of the extrajudicial "confession" Galileo would have expressed himself more or less in the same terms. Even though we are dealing with a rather partial "confession," which looked to the form rather than the content of the *Dialogue,* it provided the first admission by Galileo that he had erred. Precisely for this reason, and undoubtedly influenced by the wish to come to a "benign" solution of the trial, Maculano must have considered Galileo's confession positively, even to the point of presenting it as a true confession in his letter to Francesco Barberini. Even now, in the circumstances of a judicial confession, Maculano seemed to be satisfied with it. In fact, on that same day, April 30, the commissary of the Holy Office, "having considered the bad health and the advanced age of the above mentioned Galileo Galilei, granted that he could return to the Tuscan embassy, which would have been for him *loco carceris* (a substitute prison), with the order to deal with no one, except with those who dwelt there, to present himself at every request of the Holy Office and to keep silent under oath" (Galileo, *Opere,* 19:344). In the related document, Maculano stated that he had obtained beforehand the pope's permission. Given the fact that Urban VIII was at Castel Gandolfo, about thirty kilometers southeast of Rome, it seems utterly improbable that Maculano had succeeded on that very day of April 30 to proceed to the judicial deposition of Galileo and then to request and obtain from the pope the permission to send Galileo off to Palazzo Firenze. It is much more probable that Maculano based these actions on a previous consent of Urban VIII, which had been communicated to him by Francesco Barberini.

Niccolini gave notice of this on May 1:

> Signor Galileo was sent back to my house yesterday when I was not expecting it at all, since this examination has not been completed, and this came about through the offices of the Father Commissary together with Cardinal Barberini who on his own, without the Congregation, had him freed so that he could recover from the discomforts and ill health that plagued him continuously. Father Commissary himself also manifests the intention of wishing to arrange it that this cause be dropped and that silence be imposed on it; and if this is achieved, it will shorten everything and will free many from troubles and dangers. (Galileo, *Opere,* 15:109–10)

Niccolini was obviously well informed, almost certainly by Maculano himself. As to Maculano's intention to arrange it that the case would be annulled and no more said about it, this was an intention that, if true, was probably born from a personal wish of his and perhaps from a less rigorous interpretation by him of the theological censure to be given to the theory of Copernicus. It is possible that such an interpretation was agreed to by Francesco Barberini. But it remained at any rate subordinate to the final decision of the Holy Office. And this decision, as we shall see, would be very different.

The following letter of May 3 was also full of optimism. Galileo was already in better health, and the commissary seemed to have the intention of coming to visit him in order to bring the trial to as rapid a conclusion as possible, "by continuing the business of being as nice to him as possible and of showing himself most well inclined towards the Most Serene House." But the visit of the commissary did not come to pass. Was this a first warning of the coming change in the situation? It is hard to say. On May 10, Galileo was instead recalled to the Holy Office. From the proceedings of the trial we know that the commissary "assigned to him a period of eight days to put together his defense, if he would have the intention to do so." But Galileo had already prepared this defense, and so he delivered it immediately, together with the original document given to him by Bellarmine. After having signed the minutes, Galileo was allowed to return to the embassy. Thus far everything seemed to go on as foreseen. The fact that Galileo had his defense already prepared is an indication of it.

In his defense, Galileo explained the reasons why, at the time he requested permission to print the *Dialogue,* he had not informed Father Riccardi of the command that, according to the Holy Office version, had enjoined him "not to hold, defend, or teach in any way the Copernican opinion." To this end Galileo attached the original declaration given to him by Bellarmine (and he explained the circumstances in which he had requested this from him). It ensued from this declaration that he had been told only that the doctrine attributed to Copernicus could not be held or defended. This was the tenor itself of the decree of the Congregation of the Index, known to everyone. Galileo emphasized the fact that, relying upon this one document of Bellarmine, it was not possible to deduce that any specific command had been given him sixteen years ago and therefore that he was "quite reasonably excused" for not having thought it necessary to notify Father Riccardi of the command that the latter was familiar with (that is, the general command of the Congregation of the Index).

But what then was to be said about the personal command, with the words *vel quovis modo docere* ("or to teach in any way whatsoever"), which, according to the version of the Holy Office, had been enjoined him at the time of his summons by Bellarmine? As he had already done in the course of the first interrogation, Galileo did not deny the possibility that the order had been imposed on him, but he excused himself saying that he had forgotten, in the long intervening time, its exact formulation, so much so that the attestation of Bellarmine had freed him from any worry about other details. Galileo, therefore, showed the strong hope that his judges would remain convinced that he had not "knowingly and willingly transgressed the command" that had been given him. But he admitted again that he had been excessive in the arguments proposed in his book "because of vain ambition and the pleasure of appearing more clever than the common run of popular writers," and he said that he was prepared to correct the incriminating passages. And he ended by putting himself at the mercy of the judges to whom he made known:

> The pitiable state of ill health to which I am reduced, due to ten months of constant mental distress, and the discomforts of a long and tiresome journey in the most awful season and at the age of seventy; I feel I have lost the greater part of the years which my previous state of health promised me. (Galileo, *Opere,* 19:347; trans. Finocchiaro, *Galileo Affair,* 281)

After that final act, Galileo must have thought that he was by now close to a benign solution of the trial. Niccolini was received by Urban VIII on May 21 and came to know from him and from Cardinal Francesco Barberini that the trial would be concluded "easily" (*si terminerà facilmente*) with the "second congregation" of the Holy Office, which—Niccolini adds, writing the following day to Cioli—was "set for Thursday, eight days from today" (Galileo, *Opere,* 15:132). Now, in the year 1633, May 22 was a Sunday, so that the first Thursday "congregation" of the Holy Office would have been on May 26 and thus the "second congregation" on June 2: eleven days (not eight) after the date of Niccolini's letter. It is strange that he could have made such a mistake. One might advance the hypothesis that by the term "second congregation" the ambassador had referred to the congregation of the Holy Office to be held, rightly, eight days later, on Monday, April 30. In the ecclesiastical calendar, Monday is designated with the Latin term *feria secunda* (accordingly, Thursday becomes *feria quinta*). Even though the meetings of the *feria secunda* were restricted to the consultors, they were very important. On that occasion a review of the trial proceedings was made, and, in the case of an acknowledgment of crime, a proposal for the sentence of condemnation, including the penalties to be imposed on the culprit. In this sense, Urban VIII and Francesco Barberini could have spoken of a "conclusion" of Galileo's trial on that occasion. In favor of this hypothesis one could add that on the Thursday immediately following Niccolini's letter (May 26) a meeting of the Holy Office could not have taken place, since it was the feast-day of the *Corpus Domini*. Therefore the first Thursday congregation of the Holy Office would fall on June 2, and the "second" one on June 9, too far for the Niccolini affirmation to be correct. The more so if we pay attention also to the word used by the pope (*facilmente*), which seems to indicate an event in the very near future. I have insisted on this dating problem because, if my hypothesis is correct, one could find in the statement of Urban VIII and Francesco Barberini a signal that a "benign" outcome of Galileo's trial was still possible at this date. The consultors, on Monday (May 30), could have proposed a condemnation of the *Dialogue,* with a written retraction from Galileo and his temporary house arrest, as it had been foreseen in the instruction given by the cardinal nephew to Maculano. This seems confirmed by the comment added by Niccolini in his letter to Cioli. In fact, he considered very probable the condemnation of the *Dialogue,* unless his proposal was accepted, to let Galileo himself write an apology. And he added that "some salutary penance" would also be imposed on the latter for having transgressed the command issued to him by Bellarmine in 1616. Niccolini however had not yet mentioned this to Galileo, preferring to "prepare him by degrees, in order not to distress him."

In fact, to all appearances things continued to be encouraging. Galileo was permitted "to be able to go out of the house once in a while, to get a bit of air and to walk." And Galileo was even able to go "in a half-covered carriage" through the gardens in Rome and surrounding areas, even as far as Castel Gandolfo.

The days went on and nothing seemed to happen. In reality, the storm was brewing. In fact a decision was maturing that was far more rigorous than that which could have been hoped for from the information received by Ambassador Niccolini. According to various biographers of Galileo, that would have been due to a change in the equilibrium existing up until that time within the Holy Office between a more benign tendency towards Galileo and a more rigorous one, which in the end prevailed. The "rigorists" would only have reluctantly and momentarily accepted the proposal of Maculano. Once they had viewed together the statements made by Galileo in the course of the extrajudicial proceedings and the defense document, which he presented subsequently, they would have expressed their own strong dissatisfaction, not to say their indignation. Furthermore, according to these biographers of Galileo, the action of the "rigorists" within the Holy Office can be detected in the first part of the document *Contro Galileo Galilei* ("Against Galileo Galilei"), which summarizes the charges against Galileo, the events surrounding the permission to print the *Dialogue,* and the content of the interrogations (including the defense presented by Galileo). This trial summary was most likely composed—as customary—by the assessor, Pietro Paolo Febei.

At the beginning of it there was placed the denunciation of the *Letter to Castelli,* made by Lorini in February 1615, and the incriminating passages from it were reproduced according to the text sent by Lorini himself, but without mentioning the nonadverse censure that the work had received. Then there was placed the interrogation of Caccini of March 20, 1615:

> Father Caccini was examined, and, besides the above-mentioned matters, he testified having heard Galileo utter other erroneous opinions: that God is an accident; that He really laughs, cries, etc.; and that the miracles attributed to the Saints are not true miracles. (Galileo, *Opere,* 19:293; trans. Finocchiaro, *Galileo Affair,* 282)

But in reality not even Caccini himself had stated that such opinions had been expressed by Galileo. It is true that immediately afterward the summary specified that from an examination of the witnesses "it is deduced that such propositions were not assertive on the part of Galileo and his pupils, but only putative." But even this is false. In the deposition of Attavanti these propositions are, in fact, not attributed in any way, even in a putative form, to Galileo.

Undoubtedly this part of the document seems to have been composed with the intention of putting Galileo in a bad light. Such an intention is also clear at times in the following parts of the document. Thus, for example, it reports with respect to the book on the sunspots:

> Then, from the book on sunspots published in Rome by the same Galileo, two propositions were examined: "that the sun is the center of the world and wholly motionless regarding local motion; that the earth is not the center of the world and moves as a whole and also with diurnal motion." They were qualified as philosophically absurd. Moreover, the first was also qualified as heretical, for expressly conflicting with Scripture and the opinion of the Saints; the second as at least erroneous in faith, considering the true theology. (Galileo, *Opere,* 19:294)

Now, as already remarked, these propositions are not at all found in that book of Galileo but instead are taken from the deposition of Caccini, even though it is true that they substantially reflect Galileo's position. The inexactness of the summary is also evident with respect to Bellarmine's precept to Galileo. The papal decision is resumed as follows:

> Consequently on February 25, 1616, His Holiness ordered the Lord Cardinal Bellarmine to summon Galileo and give him the injunction that he must abandon and not discuss in any way the above mentioned opinion of the immobility of the sun and the motion of the earth. (Galileo, *Opere,* 19:294)

As we know, the papal order had foreseen two possible phases of the injunction. The compiler of the summary must have known this, since he had available to him—and clearly was using—the complete dossier of Galileo, as shown by the references to it included in the document. Even more inexact is the following statement of the document:

> On the 26th the same Cardinal, in the presence of the Father Commissary of the Holy Office, notary, and witnesses, gave him the said

injunction, which he promised to obey. Its tenor is that "he should abandon completely the said opinion, and indeed that he should not hold, teach, or defend it in any way whatever; otherwise the Holy Office would start proceedings against him." (Galileo, *Opere,* 19:294; trans. Finocchiaro, *Galileo Affair,* 282)

As we see, the injunction is quoted correctly, but it is attributed to Bellarmine and not to Segizzi. During the interrogations and in his written defense, Galileo had always insisted that the only person from whom he had received an injunction not to hold or defend the Copernican opinion was Bellarmine. He had also admitted the possibility that the cardinal may have added the words "not to teach" and "in any way whatsoever," but at the same time he had invoked a lack of memory, given the length of the elapsed time and the absence of those words in the certificate he had obtained from Bellarmine. The compiler of the trial summary seems to have wanted to make use of this version of what happened so as to do away with the problem of the apparent irregular intervention of Segizzi. In fact, Commissary Segizzi is recorded here as being present but without intervening. In doing so, however, the trial summary ends by making Bellarmine culpable of not having followed the instructions, which had foreseen an intervention with those words by the commissary only if and after the precept in a milder form had been administered to Galileo by the cardinal.

As to the testimonial of Bellarmine, the summary could not avoid citing it in its summary of the first interrogation: "Galileo confesses the order, but on the basis of the said fact [of the testimonial of Bellarmine] in which are not recorded the words not to teach in any way whatsoever, he says that of this he has no recall" (Galileo, *Opere,* 19:295). As we see, a seemingly complete confession of the injunction is now attributed to Galileo, with the result that the testimonial of Bellarmine is neutralized. In fact, if the complete injunction had been given by Bellarmine, with all of the expressions "confessed" by Galileo himself, the missing expressions "not to teach" and "in any way whatever" in the testimonial become insignificant.

The summary was most probably submitted to the consultors of the Holy Office during their meeting of the *feria secunda* (Monday), for the final assessment of the facts concerning Galileo's trial, and consequently for the proposal of the final sentence, to be submitted to the cardinals of the Holy Office and to the pope. If one accepts the already mentioned hypothesis that such meeting took place on Monday, May 30, the question remains of the delay of seventeen days for the final decision taken by Urban VIII on June 16. Here, we do not have any documentary justification for this fact. Any attempt to explain it cannot pretend to offer more than a sheer guess. With this caveat, let us imagine that having viewed the content of the summary, the consultors decided that the matter was a very serious one, and thus they preferred to leave the final decision to the cardinals of the Holy Office and the pope himself. On the following June 1, at the Wednesday meeting of the cardinals, there was a lively debate on what to do, in view of the content of the summary. On the one hand, through the reading of the summary, the cardinals had come to know the misrepresentations concerning Galileo without the possibility of knowing them as such. At the same time, they had became aware of the existence of the qualification of the Copernican thesis as—at least—an error in faith, made on February 1616. On the other hand, they must have concluded that the "confession" made by Galileo as a result of the extrajudicial initiative of Maculano was a far cry from a truthful one. If some of the cardinals, such as Bentivoglio, had been until then in favor of a benign solution of the trial, they must have felt that a dramatic change in the situation had occurred, making such a solution impossible. Supposing that Francesco Barberini had been present at the meeting, one could further guess that later on he might have reported on the new situation to Urban VIII and that the pope decided to take more time for his final decision, thus postponing it for two weeks. With this not altogether impossible scenario one could explain why the early conclusion of the Galileo trial, foreseen by the pope and Francesco Barberini on May 21, failed to materialize.

Even supposing that up to that time Urban VIII had not yet arrived at a decision concerning his old "friend," the fact is that on the occasion of the meeting of the Holy Office, held on Thursday, June 16, that decision had finally been taken. Here is the text of the acts of the session held on that day:

> The case of the Florentine Galileo Galilei having been put forth
> and a report having been given on the trial etc., and the votes hav-

ing been heard, *Sanctissimus* [the pope] decreed that the said Galileo
is to be interrogated on his intention, even with the threat of torture; and this having been done [*si sustinuerit*] he is to abjure under
vehement suspicion of heresy [*de vehementi*] in a plenary session of
the Congregation of the Holy Office; then is to be condemned to
the imprisonment at the pleasure of the Holy Congregation, and
ordered not to treat further, in whatever manner, either in words
or in writing, on the mobility of the Earth and the stability of the
Sun or against it; otherwise he will incur the penalties of relapses.
The book entitled *Dialogo di Galileo Galilei Linceo* is to be prohibited.
(Galileo, *Opere*, 19:283; trans. Santillana, *The Crime of Galileo*, 292–93,
with slight modifications)

The big question mark in the whole Galileo trial concerns the reason (or better, reasons) why this final decision of the pope was much
more severe than the previous happenings had led to foresee. Here, too,
one can only guess. A first reason may have been Urban VIII becoming
aware of the content of the summary, certainly after May 21. Through
it, he was able to find out what kind of "confession" Galileo had made
at the time of the extrajudicial initiative of Maculano. It is possible that
the pope felt in it a new attempt at a *raggiro* made by his old "friend,"
a final challenge to his pontifical authority. The deep grudge he had
borne in his heart for almost a year could have exploded now into real
indignation. Once more he must have felt betrayed by his old "friend."
If up to that very moment Urban VIII had possibly remained undecided
about the final decision to be taken, now no doubt was left in his mind.
Galileo had to learn once for all that it was inadmissible to ridicule the
authority of the Church and of that of his old patron, the pope.

The summary, again, gave him the opportunity for revenge. It had
reported—as we know—the answers given in 1616 by the qualifiers
of the Holy Office. According to them, the two assertions on the Copernican theory attributed to Galileo, taken together, implied an "error
in faith." The decree of March 5, 1616, published by the Congregation
of the Index, had avoided taking on this qualification and had simply
defined heliocentrism as a "false Pythagorean teaching" and completely
contrary to Holy Scripture. This ambiguous formulation left open two
different theological interpretations. The first was "benign": it was

"rash" to uphold the theory of Copernicus, and Urban VIII himself— we know—had supported this interpretation in the past. Anyone guilty of doing so would have to abandon his thinking, but without having to abjure. This was probably also the interpretation of Maculano, as well as of Francesco Barberini, during the Galileo trial. The second interpretation was rigorous: even if not formal heresy, to uphold the theory was at least an "error in faith." Anyone guilty would be judged "vehemently suspect of heresy" and, as such, would be obliged to make an abjuration and would be condemned to the prison of the Holy Office.

Urban VIII appears to have decided to use his magisterial authority as the head of the Catholic Church to clarify once and for all the theological status of the question, and he decided in favor of that more severe interpretation. The decree of June 16 has—as its theological foundation—this doctrinal definition of the pope.

As a supplementary but no less weighty reason for such a decision, there was no doubt the will to show to the cardinals Urban VIII's zeal for the cause of religion, which had been openly challenged in such a harsh way—one year before—by one of them, the cardinal Borgia. By condemning even the "universally" famous Galileo, his old "friend," to a humiliating abjuration, the pope wanted to make very clear his concern for orthodoxy, even to the point of including the terrible threat of burning at the stake should Galileo commit another transgression (*sub poena relapsus*).

Last, but not least, by the severity of this condemnation Urban VIII wanted to eliminate once and for all any suspicion that weighed upon him of having been the one at the top responsible for the granting of the *imprimatur* of the *Dialogue* in Rome. But at the bottom of all such severity was no doubt his personal resentment for having been betrayed by Galileo. This was an infraction against the fundamental rule of patronage and it would never be pardoned.

According to the decree, the final decision in Galileo's trial was taken upon "the votes having been heard" of those present (namely, six of the ten cardinal members of the Holy Office). No doubt, however, that decision took place as the result of a personal decision by the pope more than because of a final prevalence of the rigorist faction of the cardinals.

When Niccolini appeared three days later in an audience with Urban VIII, he was received by him with "an infinity of most kind demonstrations." Encouraged, Niccolini begged yet once again that Galileo's cause be resolved with dispatch. The pope responded that it had already been concluded and that Galileo would be called to the Holy Office to hear the sentence some morning of the following week. When Niccolini pleaded that its rigor be mitigated as a gesture of benevolence towards the grand duke, Urban VIII responded that he

> had willingly done every favor to Mr. Galileo, out of the warmth he feels towards the Most Serene Patron. However, he said that in regard to the issue, there is no way of avoiding prohibiting that opinion, and it is erroneous and contrary to the Holy Scripture dictated by the mouth of God; and in regard to the person, as ordinarily and usually done, he would have to remain imprisoned here for some time, because he disobeyed the orders he received in the year 1616; but as soon as the sentence is published, His Holiness will see me again and will discuss with me what can be done to cause the least pain and the least affliction to him, for there is no way of avoiding some personal punishment. (Galileo, *Opere,* 15:160; trans. Finocchiaro, *Galileo Affair,* 259)

To Niccolini's insistence, as he came back to his plea for clemency "towards the advanced age of seventy years of this good old man and towards his sincerity," the pope repeated that it would not have been possible not to take some measure of confinement against Galileo, since the Congregation "all of it united and no one dissenting was inclined in this direction about penalizing him." As usual, Niccolini limited himself to reporting to Galileo only the part of the conversation less difficult to bear by informing him of the proximate conclusion of the trial and of the prohibition of the book, without mentioning the punishment to be provided against him.

Galileo was not, therefore, prepared except partially for that which awaited him when he was requested to present himself to the Holy Office on the morning of Tuesday, June 21. The interrogation began, according to set plans, with the question put to him by Commissary

Maculano *super intentione,* with the purpose of clarifying once for all the real intention that Galileo had had in writing the *Dialogue.* To the question as to "whether he holds or has held, and how long ago, that the Sun is the center of the world and that the Earth is not the center of the world and moves, and also with a diurnal motion," he answered:

> A long time ago, that is, before the decision of the Holy Congregation of the Index, and before I was issued that injunction, I was undecided and regarded the two opinions, those of Ptolemy and Copernicus, as disputable, because either the one or the other could be true in nature. But after the above-mentioned decision, assured by the prudence of the authorities, all my uncertainty stopped, and I held, as I still hold, as very true and undoubted Ptolemy's opinion, namely the stability of the earth and the motion of the sun. (Galileo, *Opere,* 19:361; trans. Finocchiaro, *Galileo Affair,* 286)

Maculano objected that from the manner and content in which the said opinion was discussed in the book, and from the very fact of his having written and printed the said book, he was presumed to have held that opinion after the time specified, and urged Galileo to freely tell the truth whether he holds or had held the same opinion. Galileo answered:

> In regard to my writing of the *Dialogue* already published, I did not do so because I held Copernicus's opinion to be true. Instead, deeming only to be doing a beneficial service, I explained the physical and astronomical reasons that can be advanced for one side and for the other; I tried to show that none of these, neither those in favor of this opinion or that, had the strength of a conclusive proof and that therefore to proceed with certainty one had to resort to the determination of more subtle doctrines, as one can see in many places in the *Dialogue.* So for my part I conclude that I do not hold and, after the determination of the authorities, I have not held the condemned opinion. (Galileo, *Opere,* 19:361–62; trans. Finocchiaro, *Galileo Affair,* 287)

As one can see, Galileo based his pretense of not holding the Copernican thesis upon "superior motives." It was the Urban VIII thesis, which Galileo was trying once more to present as his own. But the interrogator was not convinced, retorting that:

> From the book itself and the reasons advanced for the affirmative side, namely that the earth moves and the sun is motionless, he is presumed, as it was stated, that he holds Copernicus's opinion, or at least that he held it at the time, therefore he was told that unless he decided to proffer the truth, one would have recourse to the remedies of the law and to appropriate steps against him. (Galileo, *Opere,* 19:362; trans. Finocchiaro, *Galileo Affair,* 287)

Galileo however was resolute:

> I do not hold this opinion of Copernicus, and I have not held it after being ordered by injunction to abandon it. For the rest, here I am in your hands; do as you please. (Galileo, *Opere,* 19:362; trans. Finocchiaro, Galileo Affair, 287)

Father Maculano exhorted him once more to tell the truth, because "they will otherwise have recourse to torture." But Galileo had no fear, nor did he change his tone. "I am here to obey (*per far l'obbedienza*), but I have not held this opinion after, as I said." The minutes end with the comment "since nothing else could be done for the execution of the decision, after he signed he was sent to his place," that is, to his dwelling in the Holy Office.

The minutes of this interrogation are sufficient to prove that Maculano limited himself to the *territio verbalis,* that is, to a purely verbal threat of torture, without coming right to the *territio realis,* by showing the instruments of torture, and even less to the torture itself. In the contrary case, the minutes should have reported the *territio realis* as is always done in similar cases. And in the case of torture, there would have to be reported also the words pronounced by Galileo under its application. The great majority of the recent authors agree on this point. On the other hand, this rigorous examination *super intentione* was little more than a formality. That Galileo had violated the precept of 1616 was already certain from the report of the theologians who had examined the *Dialogue.* He was already, objectively, vehemently suspected of heresy. What the Holy Office wanted to make sure of, in fact, through the rigorous examination, was his disposition to "be obedient" (*far l'obbedienza*). As a matter of fact, the sentence of condemnation itself will recognize that Galileo, in the course of this rigorous examination, had answered "in a Catholic way," showing his disposition to obey.

Galileo was kept at the Holy Office until the following day. We can well imagine what must have been the state of mind with which he passed those long hours. He was by now aware of the failure of all of his efforts in favor of the Copernican thesis. And, in contrast to 1616, this time his book and he himself would be at the center of the Church's decision. As to the *Dialogue,* Galileo knew already that it would be prohibited. He also knew that severe measures of penance awaited him. But did he foresee the abjuration? We do not know. At any rate, final illusions, if there were any, were completely extinguished when he was forced to put on the white penitential garb and was led to the Dominican convent of Santa Maria sopra Minerva, at the center of Rome. Kneeling down in the presence of the cardinals and of the officials of the Holy Office, Galileo heard the reading of the sentence of condemnation:

> We, Gasparo Borgia, with the title of the Holy Cross in Jerusalem; Fra Felice Centini, with the title of Santa Anastasia, called d'Ascoli; Guido Bentivoglio, with the title of Santa Maria del Popolo; Fra Desiderio Scaglia, with the title of San Carlo, called di Cremona; Fra Antonio Barberini, called di Sant'Onofrio; Laudivio Zacchia, with the title of San Pietro in Vincoli, called di San Sisto; Berlinghiero Gessi, with the title of Sant'Agostino; Fabrizio Verospi, with the title of San Lorenzo in Panisperna, of the order of priests; Francesco Barberini, with the title of San Lorenzo in Damaso; Marzio Ginetti, with the title of Santa Maria Nuova, in the order of deacons;
>
> By the grace of God, Cardinals of the Holy Roman Church, and especially commissioned by the Holy Apostolic See as Inquisitors-General against heretical depravity in all of Christendom, . . .
>
> We say, pronounce, sentence, and declare that you, the above-mentioned Galileo, because of the things deduced in the trial and confessed by you as above, have rendered yourself according to

this Holy Office vehemently suspected of heresy, namely of having held and believed a doctrine which is false and contrary to the divine and Holy Scripture: that the sun is the center of the world and does not move from east to west, and the earth moves and is not the center of the world, and that one may hold and defend as probable an opinion after it has been declared and defined contrary to Holy Scripture. Consequently you have incurred all the censures and penalties imposed and promulgated by the sacred canons and all particular and general laws against such delinquents. We are willing to absolve you from them provided that first, with a sincere heart and unfeigned faith, in front of us you abjure, curse, and detest the above-mentioned errors and heresies, and every other error and heresy contrary to the Catholic and Apostolic Church, in the manner and form we will prescribe to you.

Furthermore, so that this serious and pernicious error and transgression of yours does not remain completely unpunished, and so that you will be more cautious in the future and an example for others to abstain from similar crimes, we order that the book *Dialogue* by Galileo Galilei be prohibited by public edict.

We condemn you to formal imprisonment in this Holy Office at our pleasure. As a salutary penance we impose on you to recite the seven penitential Psalms once a week for the next three years. And we reserve the authority to moderate, change, or condone wholly or in part the above-mentioned penalties and penances.

This we say, pronounce, sentence, declare, order, and reserve by this or any other better manner or form that we reasonably can or shall think of.

So we the undersigned Cardinals pronounce:

    Felice Cardinal d'Ascoli.

    Guido Cardinal Bentivoglio.

    Fra Desiderio Cardinal di Cremona.

    Fra Antonio Cardinal di Sant'Onofrio.

    Berlingiero Cardinal Gessi.

    Fabrizio Cardinal Verospi.

    Marzio Cardinal Ginetti.

(Galileo, *Opere,* 19:402–6; trans. Finocchiaro, *Galileo Affair,* 289–91)

After the reading of the sentence for Galileo there was nothing more except to obey. Still kneeling down he read the formula of abjuration that had been presented to him:

I, Galileo, son of the late Vincenzio Galilei of Florence, seventy years of age, arraigned personally for judgment, kneeling before you Most Eminent and Most Reverend Cardinals Inquisitors-General against heretical depravity in all of Christendom, having before my eyes and touching with my hands the Holy Gospels, swear that I have always believed, I believe now, and with God's help I will believe in the future all that the Holy and Apostolic Church holds, preaches, and teaches. However, whereas, after having been judicially instructed with injunction by the Holy Office to abandon completely the false opinion that the sun is the center of the world and does not move and the earth is not the center of the world and moves, and not to hold, defend, or teach this false doctrine in any way whatever, orally or in writing; and after having been notified that this doctrine is contrary to Holy Scripture; I wrote and published a book in which I treat of this already condemned doctrine and adduce very effective reasons in its favor, without refuting them in any way; therefore, I have been judged vehemently suspected of heresy, namely of having held and believed that the sun is the center of the world and motionless and the earth is not the center and moves.

Therefore, desiring to remove from the minds of Your Eminences and every faithful Christian this vehement suspicion, rightly conceived against me, with a sincere heart and unfeigned faith I abjure, curse, and detest the above-mentioned errors and heresies, and in general each and every other error, heresy, and sect contrary to the Holy Church; and I swear that in the future I will never again say or assert, orally or in writing, anything which might cause a similar suspicion about me; on the contrary, if I should come to know any heretic or anyone suspected of heresy, I will denounce him to this Holy Office, or to the Inquisitor or Ordinary of the place where I happen to be.

Furthermore, I swear and promise to comply with and observe completely all the penances which have been or will be imposed upon me by this Holy Office; and should I fail to keep any of these promises and oaths, which God forbid, I submit myself to all the penalties and punishments imposed and promulgated by the sacred canons and other particular and general laws against similar delinquents. So help me God and these Holy Gospels of His, which I touch with my hands.

I, the above-mentioned Galileo Galilei, have abjured, sworn, promised, and obliged myself as above; and in witness of the truth, I have signed with my own hand the present document of abjuration and have recited it word for word in Rome, at the convent of the Minerva, this twenty-second day of June 1633.

I, Galileo Galilei, have abjured as above, by my own hand. (Galileo, *Opere,* 19:406–7; trans. Finocchiaro, *Galileo Affair,* 292–93)

It is to be noted that of the ten cardinal members of the Holy Office, only seven were present and signed the sentence. Among the absentees was Francesco Barberini, who also did not take part in the meeting of June 16. Many Galileo biographers have seen in this absence a sign of his disapproval of Galileo's condemnation. This is possible. As to his absence at the moment of the abjuration, however, it is to be remarked that on that same day he had to participate, together with Cardinal Borgia (another absentee), in a meeting with the pope, as it is evident from a letter of his, written on the following day. And this could have prevented him from assisting at the abjuration.

As is evident from the long motivation of the sentence, which precedes the latter and has been here omitted, Galileo is condemned both for having transgressed the decree of the Index (in which the Copernican doctrine is declared "false and altogether contrary to divine Scripture") and for having "artfully and cunningly" extorted the printing permission for the *Dialogue,* without manifesting to the ecclesiastical authorities the existence of the precept and of the injunction. As to the first transgression, the sentence notices that Galileo's pretense of having left "undecided and in express terms as probable" the question of Copernicanism is "a most grievous error, since an opinion can in no wise be probable which has been declared and defined to be contrary to divine Scripture." As to the second transgression, the sentence remarks that even allowing that Galileo, at the moment of asking for the print-ing permission of the *Dialogue,* had lost memory of the rigorous precept of Segizzi, and kept only that of the one "delicately" communicated to him by Bellarmine, witness of which was the declaration released by the cardinal to Galileo, the existence of the latter was not an extenuating but, on the contrary, an aggravating circumstance for the accused "since, although is there stated that the said opinion is contrary to Holy Scrip-ture," he had "nevertheless dared to discuss and defend it and to argue its probability."

To be sure, Galileo had always denied, in the course of the trial and even at the moment of the examination *super intentione,* any intention to defend the Copernican system or even only to have presented it as probable. But the affirmations, on the contrary, of the sentence appear sufficiently objective to us, no less than to the Galileo judges. They, therefore, could not but pronounce a sentence of condemnation.

As to its theological motivation, namely, "for a vehement suspi-cion of heresy," it was, as already noted, the result of a doctrinal deci-sion of Urban VIII, with which the pope had made his own the theo-logical qualification of the Copernican thesis as (at least) "erroneous in faith." Once such vehement suspicion of heresy had been ascertained, the only means offered to the culprit to "purge" himself from that sus-picion was to abjure it. A refusal to abjure would transform that suspi-cion of heresy into a "proven" heresy, and the inevitable penalty in this case was to be burned at the stake. This explains how, according to the inquisitorial norms of the time, the abjuration was the normal (and happy!) conclusion of a trial at whose end the accused (during the ex-amination *super intentione*) had answered, as Galileo had done, "catholi-cally," namely, had shown himself ready to be obedient (*far l'obbedienza*) and to renounce his errors. The abjuration was indeed the juridical proof of such a renunciation.

Together with this function of purging from a suspicion of heresy, however, the abjuration constituted also a most grave precedent in the case of a subsequent relapse. In such a case, a relapse into heresy, the only possible outcome was burning at a stake, even in the event of a new repentance, with the only difference—in the latter case—of the hang-ing of the culprit and subsequent burning at the stake of his cadaver.

Thus Galileo's judges could be certain that with his condemnation for a vehement suspicion of heresy he would have been silenced forever, as, in fact, was the case. To make an overall assessment of the trial, it must first of all be admitted that in the course of it Galileo was treated with respect altogether exceptional in the practice of the time. Speaking with Ambassador Niccolini, Urban VIII stressed several times, however, that such special treatment was only a sign of respect towards the grand duke. As to the juridical aspect of the trial, a first question concerns the authenticity of the document with the Segizzi injunction. The elements in favor of such authenticity appear now to possess an overwhelming weight, in spite of a very recent attempt to prove again its falsity (on the line of Santillana's thesis) by Vittorio Frajese. The hypothesis of a false document, moreover, implies an extremely grave judgment on the practice of the Roman Inquisition. In order to endorse it, one would have to have sufficiently valid and weighty reasons, which seem to be lacking altogether. After all, even concerning the actions of the Holy Office, the fundamental principle of Roman Law, *nemo reus nisi probetur* ("nobody is culpable unless so proven"), is to be considered valid.

On the other hand, the weight of this document on the conclusion of the trial must be reexamined. As is evident from the sentence of condemnation itself, the determining factor was that Galileo in the *Dialogue* had defended the Copernican theory. This was a violation of the decree of 1616 as well as the precept of Bellarmine (about which there did not exist, nor even now do there exist, any doubts). This would have held true in any case, even without the discovery of the document with the Segizzi injunction. The importance given to the latter seems after all to have been aimed at further exculpating the master of the Sacred Palace, and above all Urban VIII himself, from the charge of having granted the *imprimatur* for the *Dialogue*. At the moment of asking for it, Galileo had been "fraudulently" silent not only about the precept of Bellarmine but also about the existence of that injunction, and this fully exonerated Riccardi from the responsibility of having granted the *imprimatur*.

As to the serious inaccuracies and distortions of facts present in the trial summary, they could put in question the legitimacy of the condemnation only if they would have been reported during the sentence of condemnation as determining factors. This is, however, not the case. On the contrary, the sentence rectifies the quotations of the documents made by the summary and in general appears to have been written without ill-will towards Galileo. It could be, therefore, that on the occasion of the meeting of June 16, if not before it, some cardinal or consultor put in evidence such shortcomings of the summary, with the result that they were corrected. Or perhaps such a correction may have been made at the moment of the writing of the sentence by its compiler (most probably Maculano). As to the important affirmation of the sentence, taken as it was from the summary, that Galileo confessed to having received the injunction in its severe form (which the sentence correctly attributes to Segizzi), such affirmation was based on the declarations made by Galileo himself at the end of the first interrogation and in his written defense. They appear, indeed, as an acknowledgment of the fact of the injunction, even though Galileo continued to affirm, even in this defense, that its author had been Bellarmine. However, the compiler of the sentence had before his own eyes the document in which the injunction appeared to have been made by Segizzi. And he did not doubt its authenticity. Confessed or not by Galileo, the evidence of the facts was there. On the other hand, I insist once more, the evidence that Galileo, in writing the *Dialogue,* had defended the Copernican thesis was the reason that by itself alone was sufficient to justify the sentence of condemnation.

It seems therefore fair to affirm that the above deficiencies of form are not sufficient to put in doubt the juridical validity of Galileo's condemnation, at least from the point of view of the jurisprudence of his time. The judgment, obviously, is completely different from our point of view. The condemnation and the abjuration appear to us as the outcome of an abuse of power, both on the doctrinal level and with respect to Galileo's personal conscience. No doubt, one must avoid the error of projecting our modern ideal of freedom of thought back to an age of increasing absolutism, as in Galileo's time, where the principle of authority (both in civil and ecclesiastical affairs and no less in the Protestant than in the Catholic camps) was considered superior to the principle of individual intellectual freedom. It was precisely this principle of authority that the appearance of the *Dialogue* had put under scrutiny. Convinced that the "marvelous conjuncture" of the erection to the papacy of his patron and admirer, Maffeo Barberini, would not happen again and overestimating the pope's esteem for him, Galileo had in substance decided to play his cards, having been encouraged, as we have seen, by Urban VIII's apparently positive reaction to the *Letter to Ingoli.* But it was a card game that ended in disaster. Despite the rhetorical expedients that Galileo devised to maintain the purely hypothetical character of the principal conclusions in the *Dialogue,* it was, in fact, too obviously pro-Copernican to escape becoming a challenge to ecclesiastical authority and especially to the authority of Urban VIII. The challenge was all the more scandalous given the reputation of Galileo throughout Europe. It is precisely this that Urban VIII emphasized when he accused Galileo of "having given such universal scandal to Christianity," an accusation that he will subsequently continue to repeat. If one takes into account, furthermore, the very difficult political situation of Urban VIII, in that most dramatic moment of the Thirty Years War, and the criticism voiced by some of the highest members of the ecclesiastical hierarchy for his lack of "apostolic zeal," a severe reaction with respect to the publishing of the *Dialogue* was inevitable. The more so if one adds to all this the personal resentment of a patron who felt betrayed by his protégé.

One must also take into account the different juridical position in which Galileo had come to find himself in 1633, as compared to the one in 1615–16. Because of the lack of an official position of the Church concerning the Copernican theory, before February 1616, Galileo had not been personally condemned but only privately admonished by Bellarmine to abandon it. The existence of this admonition (even apart from the injunction administered abusively—as it appears—by Segizzi) and the subsequent decree of the Index are the new facts, with respect to the first indictment of Galileo, that put his trial in 1633 on a plane of an altogether different juridical nature. Galileo, no doubt, was aware of it. He, however, underestimated the risk he was taking in writing the *Dialogue,* both because of his overconfidence in the special "friendship" shown to him by the pope and because of his knowledge—through the report he obtained by Cardinal Zollern—that Urban VIII considered the holding of the Copernican thesis only as "rash." Of course, he was unable to foresee the change of mind of his old patron, with his final doctrinal decision concerning the theological qualification of the Copernican doctrine.

That decision was the juridical basis of the severity of Galileo's condemnation. Urban VIII—by his pontifical authority—had the power to make that doctrinal clarification regarding the ambiguity of the decree of 1616. The fact that he did it, and the reasons for doing it, cast a dark shadow if not on the inevitability of that condemnation, then on the legitimacy of its severity, which appears to us as the result of a blind authoritarianism. More fundamental, however, is the fact that a previous doctrinal stiffening and a similar authoritarianism are present in the doctrinal and disciplinary decisions of 1616. Such decisions, and particularly the decree of the Index, were the outcome of the preoccupation with preserving the status quo of the traditional Christian vision of the world. By insisting on such a worldview, the ecclesiastical authorities showed themselves incapable of grasping the importance of the new problematic created by the Copernican view and by Galileo's discoveries. To declare that such view was "false and altogether opposed to Sacred Scripture" meant that one wanted to cut off the problem definitively and so preclude any position of prudent waiting in view of possible future proofs. And it would seem important to notice once more that such a possibility was excluded (for the present and for the future) by the qualification of "absurd in philosophy" given by the qualifiers of the Holy Office about the Copernican theory. It was this very philosophical certitude that set up the basis for their certainty in responding theologically. For that reason, neither in 1616 nor in 1633 was any consideration given to the question as to whether Galileo did or did not possess convincing proofs for heliocentrism. In their blind adhesion to the Aristotelian view of the world, any such proof was excluded a priori.

It is this myopic authoritarianism from which flowed an undue doctrinal and disciplinary closure to a question that should have been left open, that can be considered—in fact must be considered—the original abuse of power. From it there came about, as a further hardening of position, the condemnation of Galileo.

But even more at the roots of the condemnation, the fact that it happened and the form it took can be traced to the existence of a doctrinal and disciplinary power exercised by an institution, the Holy Inquisition, whose central organ was the Holy Office. The fact that from the Middle Ages until well beyond Galileo's time there had developed in the Church a system that controlled "Catholic" thought and activity and that was, in fact, despite the attribution of "holy," based on coercion and, when necessary, on physical or psychological violence, constitutes an institutionalized abuse of power that can never be sufficiently deplored. Galileo, with his tactical errors, must undoubtedly bear part of the responsibility for the *fact* of the condemnation. But the responsibility for the *way* in which the condemnation occurred, and especially for the abjuration, falls without a doubt on the shoulders of the Church of those times and specifically on the organs and the methods that were used in the exercise of the Church's authority.

Galileo's judges surely did not have such a feeling when they returned to their homes after the congregation's session. On the contrary, they must have thought with relief that finally that thorny question had been brought to an end and that from that time on there would be no more talk of Copernicus, nor even of Galileo, at least among Catholics. Those who were more benevolent towards Galileo must, of course, have felt sorry for that poor old man, but in conscience they felt at peace. In the end, that which happened was the fault of Galileo alone, of his imprudence, obstinacy, and lack of sincerity. And there was no doubt that the Church had been magnanimous with him. All knew that imprisonment at the Holy Office was a formality on which there would be no insistence. For sure, the abjuration must have been difficult for Galileo, they would admit. But it was a lesson that in the end would save him from worse ills.

And so, for the cardinals and officials of the Holy Office, the "Galileo Case" was by now a closed chapter. They had no suspicion that the true Galileo Affair was instead just beginning on that very day of June 22, 1633, and that their names, together with those of their predecessors of 1616, would pass on to posterity not only as judges of the tribunal of the Holy Office but also and above all as the accused, destined to be called innumerable times in the centuries to come before the much more severe tribunal of history.

It would not be correct to unload their responsibilities onto Galileo's enemies, who had begun to take action against the *Dialogue* by influencing Urban VIII and continuing their activities through the faction averse to Galileo within the Holy Office. Let us begin with those who are mainly accused in this plot to ruin Galileo, that is, the Jesuits. Undoubtedly, Galileo himself was convinced that they had been at the origin of his misadventures. In this sense he had written to Diodati on the verge of his departure for Rome. Galileo's persuasion was probably based on the letter sent by Magalotti to Guiducci in August of the previous year. The statement of Magalotti, "The Jesuit Fathers must be working underhandedly and valiantly to see that the work is prohibited," depended in turn explicitly on what Riccardi had told him: "The Jesuits will persecute him in a most bitter way." This is a matter of a prediction and not of an accomplished fact.

Even a year after his condemnation, upon writing again to Diodati in Paris, Galileo showed himself convinced of the responsibility of the Jesuits in the whole drama. After having stated that the anger of his "most powerful persecutors," far from quieting down, was continuously increasing, he added:

> These persons [the persecutors] have finally wanted to reveal themselves to me, seeing that a dear friend of mine while he was in Rome about two months ago conversing with Father Christopher Grienberger, Jesuit Mathematician of that College, my affairs having come up for discussion, the Jesuit spoke these formal words to that friend: "If Galileo had known how to keep the affection of the Fathers of this College, he would live gloriously in this world and none of his bad times would have come to pass and he would have been able to write as he wished about everything, even, I say, about the motion of the Earth etc." Thus Your Lord sees that it is not this or that opinion which has caused and does cause war for me, but the fact that I am in the disfavor of the Jesuits. (Galileo, *Opere*, 16:116–17)

It is difficult to give to these statements of Galileo, based in their turn on those of a "dear friend" of his, the probative value attributed to them by many biographers of Galileo. According to the words attributed to Grienberger, he would have wanted to lead his interlocutor to believe that the mighty power of the Jesuits would have allowed Galileo (if he had remained their friend) to assert any theory, even that of the Earth's motion. Such an affirmation, in sharp contrast with the temperament and posture always shown by Grienberger, was absurd. Though the Jesuits were powerful and influential, they were certainly not omnipotent, especially with a pope like Urban VIII. Galileo (followed by many of his biographers) gave credence to that absurd and incredible remark of Grienberger, taking it as an admission from the Jesuit side that his condemnation had been their responsibility. But even supposing that Grienberger's words were authentic, they do not seem at all to imply such a responsibility. Certainly, the support of the Jesuits was not there for Galileo at the conclusive moment of his affair, as it had been lacking at the time of the Church's decision in 1616. Besides the reasons already known to us, reasons that would continue to be operative in 1633, there was undoubtedly the further reason that for some time now the relations between Galileo and the Jesuits had been irreparably ruined. Grienberger put all the responsibility for this on Galileo and that was incorrect. But in the eyes of the Jesuits the beginning of the arguments (both in the case of Scheiner and, above all, in the case of Grassi) had come from Galileo and from his friends, and in this sense Grienberger was substantially correct. However that might be, all that the by now elderly mathematician of the Roman College had stated was that Galileo, with his attitude, had lost the support of the Jesuits; not that the latter were out to persecute him.

That there were among those who attacked the *Dialogue* also some Jesuits, specifically Scheiner and Inchofer, cannot be denied. The break by now complete between the Jesuits and Galileo and the fiery polemical climate, which had already existed before the publication of the *Dialogue* and which the latter had only heated up more, could have led to hostile initiatives by one or another Jesuit, with the possibility that they had traveled even to Urban VIII. The deep hostility of the censure of the *Dialogue* composed by Inchofer during the Galileo trial, and new aspects on the role that Inchofer (a friend of Scheiner) played in the Galileo Affair, made clear by recent studies, seem to transform that possibility into a strong probability. But it does not seem that there are sufficient grounds (beyond "street rumors") to assert the existence of a collective action of the Jesuits directed towards getting Galileo condemned. Still under the trauma of the epilogue to the trial, Galileo was not able to resign himself to the idea that Urban VIII had been able to change his attitude towards him so radically. He had therefore to seek the reasons for that elsewhere, and he found them finally in the Jesuits, who thus became the scapegoat for his whole drama.

In reality, in the group who opposed Galileo there was a confluence of persons of diverse backgrounds and ideas (right up to the rigorist current within the Holy Office), members of various religious orders, representatives of the regular clergy, as well as intransigent lay persons and those who were jealous of Galileo's celebrity. But granted all of that, the fact remains that if Urban VIII had not been personally convinced that Galileo had betrayed his expectations and had contravened the promises made, helped by men who had been the pope's confidants, and by whom the pope had felt himself tricked, he would never have yielded to outside pressures, no matter how strong. His reaction at the time of the Morandi affair, in which Galileo's enemies had tried to involve him too, is the most evident confirmation of that.

That which Galileo always refused to believe seems unquestionable, namely, the fact that at the basis of his misadventure there was, as a matter of fact, the personal reaction of the pope to the contents of the *Dialogue*. Galileo's adversaries were probably able to hasten or facilitate this reaction. But it did not grow to assume the dimensions we know except when Urban VIII became aware of how different the *Dialogue* was from the book that he had permitted Galileo to write. And it is certain that, even without any real or presumed "machination" of Jesuits, Dominicans, or what you will, once he understood this, Urban VIII would have reacted more or less in the same way.

The Burdensome Inheritance
of the Galileo Affair

Given the purpose of this book, the nine years between the end of the trial and the death of Galileo will receive only a very brief treatment. This certainly does not mean that less value is given to that period from either a human or a scientific point of view. In fact, Galileo was never as great as he was in those final years, which saw him wrestling with incredible courage and tenacity against so much physical and moral suffering, so as to leave his most lasting scientific testament, the *Discourses*.

The day after his condemnation and abjuration, the imprisonment of Galileo in the Holy Office was commuted to one in the gardens of Villa Medici, the beautiful property of the Tuscan embassy in Rome, built by Michelangelo, as a place where the Tuscan ambassadors would spend when possible a pleasant and restful time. And the following week he was granted permission to move to Siena, under house arrest in the residence of his old friend, Archbishop Piccolomini. After another six months, Urban VIII allowed Galileo "to return to his villa [at Arcetri, on the hills of Florence] to live there in solitude, without summoning anyone, or without receiving for a conversation those who might come, and that for a period of time to be decided by His Holiness" (Galileo,

Opere, 19:389). This period, in fact, lasted until his death, and Galileo would continue to experience for the rest of his years the severity of remaining under the control of the Holy Office.

The prohibition of the *Dialogue,* his condemnation, and the abjuration had been for Galileo, without a doubt, his greatest sorrows. There awaited him after his return to Arcetri a still greater suffering of a different kind, the death of his daughter, Sister Maria Celeste. During his trial and the subsequent stay in Siena, the young religious had been close to her father in a special way, with letters from which there came forth the full richness of heart, and the profound intelligence and sensitivity, of this favorite daughter of Galileo, together with a religious conviction that was not at all ostentatious or forced. The letters in which Galileo informed relatives and friends of the news of her death are a moving testimony of the depth of his sorrow and of the void that the death of his daughter had left in his life.

As to the physical sufferings, the cruelest was surely that caused by his loss of eyesight. The worsening condition of his sight and then of his complete blindness comes back time and time again in Galileo's letters from 1630 on. In the well-known letter to Diodati (December 19, 1637), the by now blind Galileo wrote:

> Alas, my lord, your dear friend and servant Galileo has for the past month become irreparably blind. Now imagine, Your Lordship, how afflicted I am as I think about that sky, that world and that universe that I with my marvelous observations and clear demonstrations had opened up hundred and thousand times more than had been commonly seen by the sages of all bygone centuries, now for me it is diminished and limited so that it is not any greater than the space I occupy. (Galileo, *Opere,* 17:237)

In spite of so many sufferings and in the impossibility of carrying on his explicitly Copernican program, Galileo had decided to put together in a definite, systematic way his studies in mechanics. He was no doubt well aware of the importance for the Copernican issue of the results that he had obtained with them. In fact, they constituted the starting point of that new natural philosophy that one day would allow the unopposed triumph of the new vision of the world and would oblige the Church to change her mind.

Spurred on by his friends and disciples old and new, such as Castelli, Bonaventura Cavalieri, and Evangelista Torricelli, Galileo succeeded in completing his work, to which he gave the title *Discourses and Mathematical Demonstrations Concerning Two New Sciences,* published in 1638. After many difficulties had been overcome, this work was printed in 1638 in Leiden, Holland, by the famous publisher Elzevier. The same editor had already published in France (Strasbourg, 1636) the Latin version of the *Dialogue,* made by Matthias Bernegger at the request of Elia Diodati and one year later also that of the *Letter to the Grand Duchess Christina,* together with its original Italian text. These two latter publications were made possible because the decree of condemnation of Galileo had not been published in France.

Upon publication, the *Discourses* quickly became widely distributed, especially in France and Germany. Numerous copies arrived also in Rome, where the Roman authorities did not oppose their sales, evidently because in the new book Galileo had not dealt with the Copernican issue. There was a long delay before Galileo had the book in his hands, and unfortunately he could not see it since he was by now completely blind. Nevertheless, he must have felt a great joy at having thus brought to completion his scientific testament.

But even after having thus reached his objective, Galileo continued to work almost up to the verge of his death, helped by the Piarist Father Settimi, as his amanuensis, and at the end by his youngest disciple, Vincenzo Viviani, who remained at his side from October 1639 until his death. Together with him, Galileo's son Vincenzio and Evangelista Torricelli witnessed Galileo's departure. Viviani gave the following account of the event:

> On Wednesday, January 8, 1641, from the Incarnation [that is, 1642 according to the Gregorian Calendar] at four o'clock in the night, at the age of seventy-seven years, ten months, and twenty days, with philosophical and Christian constancy, he rendered his soul to his Creator, sending it forth, as far as we can believe, to enjoy and admire more closely those eternal and immutable marvels, which that

soul, by means of weak devices with such eagerness and impatience, had sought to bring near to the eyes of us mortals. (Galileo, *Opere,* 19:623)

According to the usage of the epoch, four o'clock in the night corresponds to our 10 to 11 pm. As to the day of his death, the nuncio to Florence (discussed below) gives the date of January 9. The death certificate and that of burial, reproduced in the National Edition by Favaro, are not altogether in agreement.

The news of Galileo's death was given four days later to Cardinal Francesco Barberini by the nuncio in Florence, Giorgio Bolognetti, with a short and dry message, at the end of which he informed the cardinal that the grand duke, it seemed, wished to make a sumptuous tomb for him in the Church of Santa Croce, "comparable to and facing that of Michelangelo Buonarroti." Obviously, the nuncio wanted to avoid the possibility that Galileo, even after his death, could create problems for him. Francesco Barberini also received the same news from Florentine Inquisitor Muzzarelli. Answering him on January 25, the cardinal recommended to him, in the name of the pope and of the cardinals of the Holy Office, to use his "usual skill" in order that the grand duke understand that "to construct a mausoleum to one who has been given a penance by the tribunal of the Holy Inquisition and who died while the penance was still in effect" would result in a scandal for good people "and to the prejudice of his Highness's [the grand duke's] piety." At any rate, he added, on the inscription to be placed on the tomb "one should not read words which could injure the reputation of this tribunal" (Galileo, *Opere,* 18:379–80). Even the funeral oration, if made, should have been seen and approved before it was pronounced or printed.

The same recommendations against the construction of a mausoleum were expressed on the same day by Urban VIII to Ambassador Niccolini. "It would not be at all a good example to the world," he affirmed, "that His Highness would do such thing to honor someone who had given such universal scandal with a doctrine that was condemned" (18:379).

As one sees, even nine years after the condemnation of his old "friend," Urban VIII's resentment towards him had not at all lost its force. To be sure, with those words the pope wanted also to justify before Niccolini the intransigent severity of his position, which had persisted until Galileo's death. Still the fact that neither he nor Francesco Barberini nor the nuncio in Florence were capable of a single word of human feeling with respect to the deceased is something that cannot fail to strike us today. It probably also struck Niccolini. But his long years of contact with the pope must have taught him a great deal, even the wisdom of "putting off to another time" the thought of a monument to Galileo. And this was the very thing that he counseled in the following part of the letter to the secretary of the grand duke.

This advice of Niccolini and the words that the Florentine inquisitor had placed "skillfully" in the ears of the grand duke, were sufficient to divert him from his plan. And thus the body of Galileo remained "for the moment" "in a room behind the sacristy" in the Church of Santa Croce. That "for a moment" had to go on for almost one hundred years. It was, in fact, only in 1734 that the Holy Office gave permission for the construction of a mausoleum, on the condition, however, that they would be informed of the inscription that was planned to be placed there. On March 13, 1736, Galileo's remains were moved inside the church and the following year the mausoleum was finally built, in front of those of Michelangelo and Machiavelli, with money left for the purpose by Viviani. On top of the urn that contains the remains of Galileo is mounted his bust, and to each side of it there is a statue, one of which represents Astronomy and the other Geometry. There is no third statue to Philosophy, to which Galileo, as we know, would have been even more attached than to the other two. But this void is filled by the Latin inscription GALILAEUS GALILEIUS PATRIC. FLOR. ASTRONOMIAE GEOMETRIAE PHILOSOPHIAE MAXIMUS INSTAURATOR NULLI AETATIS SUAE COMPARANDUS HIC BENE QUIESCAT [Galileo Galilei, Florentine Patrician, a Great Pioneer in Astronomy, in Geometry, and in Philosophy. Incomparable to Anyone of His Time. May He Here Rest Well].

The fact that this time the Holy Office had not opposed the construction of the monument, nor the inscription that was placed there, is indubitably an indication of an evolution in the attitude of the Roman ecclesiastical authorities with respect to the Copernican theory. Isaac

Newton had by now brought to completion the program initiated by Galileo for the edification of the new natural philosophy with his great work *Mathematical Principles of Natural Philosophy* (1687). And its knowledge and influence were by now spread throughout the learned world of Europe, including Italy. This new Newtonian physics had finally given a full theoretical justification of the Copernican system, perfected upon the basis of the three laws of Kepler. Any form of geocentrism, including that of Tycho Brahe, had thus been excluded.

And then in 1728, the discovery of the phenomenon of the aberration of starlight, made by the English astronomer James Bradley, had furnished the first geometrical argument in favor of the Earth's movement about the Sun. The Italian translation of the journal the *Philosophical Transactions of the Royal Society,* in which the discovery had been reported, had made it known in Italian scientific circles in 1734. And so we have that which Bellarmine himself had admitted (even though as an altogether improbable event), that is, the necessity to reexamine the interpretation of scriptural passages regarding the motion of the Sun and the stability of the Earth. In the face of incontestable physical proofs to the contrary, this could no longer be ignored by the Roman authorities. On the other hand, there was still the decree of the Index of 1616 and the condemnation of Galileo by the Holy Office in 1633. To officially accept the Copernican view now would imply openly acknowledging a mistake on the part of the Church. And this, in the ecclesiastical atmosphere of the epoch, was simply unthinkable. One preferred, therefore, to act not directly but "behind the scenes," hoping thus to solve such an embarrassing situation without fuss, without "scandalizing" the faithful and, to be precise, especially without compromising the prestige of the Church.

The permission for the construction of the mausoleum of Galileo had been a first, quiet step in that direction. A second one was accomplished seven years later (1741) with the authorization of the first edition of the works of Galileo, which included the *Dialogue* but excluded the *Letter to the Grand Duchess Christina.* The curator of this edition was Abbot Giuseppe Toaldo, only twenty-five years old at the time, and the work was printed in 1744 at the press of the Seminary of Padua. Toaldo was the founder of the Paduan astronomical observatory, and this explains his interest and admiration for Galileo. For the text of the *Dialogue* he was able to make use of a precious copy of this work, owned by the Padua seminary, with marginal notes and autograph additions by Galileo himself.

But the conditions for the editing of the *Dialogue* had been very restrictive. At the beginning there had to be printed the sentence of condemnation of Galileo and his abjuration. As is evident from the preface, the affirmations about the motion of the Earth contained in it had to be interpreted (still!) as a "pure mathematical hypothesis." For that reason, one had "removed or reduced to a hypothetical format the marginal annotations, which were not or did not appear to be indeterminate." For the same reason, a dissertation by Father Calmet had been added, "where an explanation is given of the meaning of the passages in Sacred Scripture pertaining to this material as by common Catholic belief."

The modalities of this Paduan edition of the Galileo works are a clear indication of how hesitant and ambiguous the adjustment to the new situation remained in the first half of the eighteenth century. The Roman authorities were trying to bring about change by slow steps and without creating any fuss, despite the more open climate inaugurated by the pontificate of Benedict XIV (1740–1758), an "enlightened Pontiff," as he was defined by the *Encyclopedia* of Diderot and d'Alembert. (In tome 4 of its first edition of 1754, the *Encyclopedia* contained, in the article on Copernicus, a long passage on Galileo, in which the desire was expressed that, as had already happened in France, the error of his condemnation would also be recognized in Italy.)

It was probably through the initiative of Benedict XIV himself that in 1757, on the occasion of the preparation for a new edition of the Index of Forbidden Books, published the following year, it was decided to eliminate from it the general decree that prohibited "all the books that teach the mobility of the Earth and the immobility of the Sun." It is to be noted in this regard that in all the editions of the Index, the Copernican works individually prohibited appeared after the names of their authors and were followed by the respective decree of prohibition. Thus, in the case of Galileo: "Galilei, Galileo. *Dialogue on the Greatest Systems of the World, Ptolemaic and Copernican.* Decree of 23 August 1634." Besides, however, these individual prohibitions, the editions up to that time had reported, among the general decrees, also one extracted, with some modifications, from the last part of the decree of the Index of 1616, which concerned a prohibition of the Copernican books in general: "all other books which likewise teach the same." Obviously, one wanted in this way to spare space in the Index and at the same time to "neutralize" all the Copernican works written after those of Galileo and Kepler. It was precisely this general prohibition that was omitted in the new edition of the Index of 1758. A determining role in that omission was played by the Jesuit Pietro Lazzari, consultor of the Congregation of the Index. Lazzari had been undoubtedly influenced by one of the most influential Jesuit men of science of the epoch, Ruggero Boscovich (1711–1787), a convinced Copernican.

Even though this general decree had been taken from the last part of the decree of the Index of 1616, its omission did not imply *ipso facto* the abrogation, in the latter, of the works of Copernicus, de Zuñiga, Foscarini, Galileo, and Kepler, which remained suspended or condemned in the edition of the Index of 1758 and would still be included in that of 1819. Evidently, it had been desired to open a free path for Copernican ideas in the future without, however, repudiating the decisions of 1616 and 1633. Apart from the usual motives of "Church decorum," such an abrogation would have obliged the authorities of the Holy Office to face a task of extreme complexity from the point of view of canon law. This is, in fact, what a few years later the then prefect of the Congregation of the Index, Cardinal Galli, would affirm as an answer to the suggestion to abrogate the prohibition of the *Dialogue,* made by the famous French astronomer J. J. de Lalande (1732–1807), during his visit to Rome.

Not too much time was needed for the ambiguity of the 1757 decision to become evident. A professor of astronomy at the University of Rome, La Sapienza, the canon Giuseppe Settele (1770–1841) had published in 1818 the first volume of his treatise *Elements of Optics and Astronomy,* dedicated to optics. Having later requested permission for the publication of the second volume, dedicated to astronomy and based on the presupposition of the truth of the Copernican theory, he saw it refused by the master of the Sacred Palace, the Dominican Filippo Anfossi. As a reason for his refusal, the latter alleged the fact that the acceptance of heliocentrism as a thesis and not only as a hypothesis was not only contrary to scripture and to the Church Fathers but also to the decrees of the Congregation of the Index and of Galileo's condemnation. Since those decisions of the Church had never been abrogated, Anfossi did not feel himself authorized to grant the *imprimatur.*

With the help of the commissary of the Holy Office, the Dominican Maurizio Olivieri, Settele brought it about that the examination of the question would be passed on to the Holy Office. In his written "Observations," Olivieri had affirmed that the original Copernican theory, accepted even by Galileo, was altogether unsatisfactory from the point of view of the "natural philosophy" of the time. For instance, to have ignored that the air had a weight and, therefore, was subject to gravity, had prevented Galileo from explaining in a satisfactory way the absence of the extremely violent winds that should have resulted from the movement of rotation and revolution of the Earth. According to Olivieri, it was for such claims in natural philosophy, more than for the theological ones, that the Holy Office and the Congregation of the Index had condemned the *Dialogue,* as well as the Copernican theory itself. The subsequent progress of scientific research, with the discovery of the gravity of air made by Torricelli, and the theory of universal gravitation of Newton, together with astronomical discoveries such as that of the aberration of starlight, had freed the Copernican system from the initial inconsistencies. And it was because of this that it was now followed by practically all astronomers.

This interpretation of Olivieri was substantially taken over by the consultor of the Holy Office, Antonio Maria Grandi, who had been requested to give advice on the matter. He had concluded: "There is nothing contrary to the fact that one might defend the opinion of Copernicus on the motion of the Earth in the manner in which today it is usually defended by Catholic authors." With the decree of August 6, 1820, the cardinals of the Holy Office accepted such advice, as well as Grandi's proposal to "insinuate" to Anfossi that he no more oppose himself to the printing of the volume and to "suggest" to Settele to show that the Copernican opinion, "as it is presently defended, is no longer subject to those difficulties to which it was liable in times gone by, before those observations which were subsequently made" (Galileo, *Opere,* 19:420).

With his advice Grandi had thus answered the recommendation of the Holy Office concerning the permission of the printing of Settele's work, that is, to find a way to protect the good name of the Holy See. And in fact, what interested the Holy Office in 1820 was not the official endorsement of the Copernican theory but only to find an honorable way out of the impasse caused by the existence of an anti-Copernican decree that had never been abrogated in an epoch when Copernicanism was accepted by everybody—with the exception of a few all-out conservatives of the type of poor Anfossi. The latter gave up his opposition only after two years of fighting with the Holy Office and only after a new decree of this congregation, dated September 11, 1822, established, under threat of severe canonical penances, that neither he nor his successors could refuse to give permission for publishing works "that deal with the mobility of the Earth and the immobility of the Sun" (Galileo, *Opere,* 19:421).

With this decree of the Holy Office the official dossier regarding the Copernican question was closed. Through a real irony of history, that same Holy Office, which in 1616 had made the anti-Copernican decisions that had been the basis for the Galileo condemnation in 1633, was now threatening "punishments at its choice" against the masters of the Sacred Palace who would further oppose permission for Copernican books. That which Galileo had written in one of his copies of the *Dialogue* became thus a reality:

> Take notice, theologians, that by your wish to make a matter of faith the propositions pertaining to the motion and to the rest of the Sun and of the Earth, you expose yourselves to the danger that you may perhaps as time goes on have to condemn as heretics those who assert, the Earth stands still and the Sun changes its place; as time goes by, I say, when it would be proven by the senses or by necessary demonstrations that the Earth moves and the Sun stands still.

Profiting by the opportunity of the solution of the "Settele affair," Father Olivieri had taken the initiative of promoting the removing the works of Copernicus, de Zuñiga, Foscarini, and Galileo from the Index of Forbidden Books. But his initiative encountered the opposition of the consultors of the Holy Office, concerned about "the (bad) appearance that the Roman Curia would have made with such a decision." One preferred, that is, to wait for the new edition of the Index, with an "omission" that would attract, it was hoped, less attention. The elimination of the above works, plus those of Kepler, from the new edition of the Index was personally decided by Pope Gregory XVI on May 20, 1833, with however the explicit and significant prescription that "no judgment on the question should be added." Two years later (1835) the new edition thus emended came out. With this last, quiet act of the Holy Office one hoped to have solved once and for all the problem of Copernicanism and thus (at least implicitly) that too of Galileo.

The following decades of the nineteenth century saw, on the contrary, an even more fierce debate on the Church's past actions with respect to the Copernican vision of the world and in particular to Galileo. Such a debate had been opened in the epoch of the Enlightenment, during which those Church actions had been seen as a case emblematic of the opposition between religious obscurantism and the conclusions, fully rational and objective, of the new science. Napoleon too had been deeply influenced by these ideas of the Enlightenment. After the conquest of the Pontifical States, he had had all of the Roman archives, including those of the Holy Office, carried to Paris (1810), aiming in particular at the documentation of Galileo's trial, which he intended to have published. After Napoleon's fall, these documents of the pontifical archives were returned to the ecclesiastical Roman authorities. However, one part of the *processi* (trial documents) of the Holy Office and of the Inquisition (among which those of Bruno's trial and the original document of the Galileo condemnation and abjuration) were lost. This was due to the fact that the person in charge of the recovery of the archives, Monsignor Marino Marini, was obliged to sell a great number of those documents as paper for pulping, in Paris, in order to reduce the expedition expenses.

In spite of the attempt at a restoration, bringing Europe back to the prerevolutionary status quo and in a certain sense even to the pre-Enlightenment one, the debate on the Galileo Affair, opened by the Enlightenment, had progressively strengthened its trust. Confronting and opposing a Church increasingly less able to maintain the ability to

exert a doctrinal influence in Catholic countries, more and more substantial parts of the Church's old "faithful" were swelling the ranks of
laicism. This latter became quite often anticlericalism, that is, a movement of open warfare against what it considered as the oppression exercised for centuries by the Church in intellectual matters. As to the
Catholics who continued instead to recognize the rights and the authority of the Church, they were often aligned in defensive positions,
but not for that reason less militant than the positions of their adversaries, from whom they came to be called "clericals." In this climate of
sharp and prejudiced polemics Galileo became for his lay biographers
the symbol of the man "above all prejudices (as he had been defined
by the *French Encyclopedia*), a martyr of intellectual and religious obscurantism." Instead, in the view of the Catholic biographers, he was seen
as the one ultimately responsible for his misadventures because of his
impetuous temperament, his scornful irony, and his pretense to have
the Copernican view recognized by the Church, in spite of the lack of
real proofs in its favor.

For an objective reappraisal of the Galilean question it became
even more urgent for the Church to put at the disposal of scholars
the documents concerning Galileo, which had finally come back to the
Roman archives. But towards the middle of the nineteenth century the
tensions and the antagonisms in intellectual matters, to which were
added those in the political field caused by the movement for the unification of Italy, were too strong to allow the Church a courageous decision on the matter. As usual, external pressures made it impossible to
delay a change of attitude of ecclesiastical authorities. The first of such
pressures came about as a consequence of the creation of the ephem-
eral "Roman Republic" (1848–49) and of the concomitant flight of
Pius IX to the stronghold of Gaeta, in the territory of the kingdom of
the two Sicilys. The brief interregnum made possible the "violation"
of the Archives of the Holy Office by two members of the government
of the republic. On that occasion they hurriedly compiled copies of
documents that interested them. In particular, one of the two, Silvestro
Gherardi, transcribed from the registry of the decrees those texts that
he could find, with the intention of publishing them. He could not find,
however, the most important document, namely the volume of the

Galileo trial, which had been, in fact, entrusted by Pius IX, before he took refuge in Gaeta, to the prefect of the Secret Vatican Archives, Marino Marini. That was yet a further indication of the importance attributed by the Church to guard jealously these Galilean documents. Gherardi was thus not able to consult them. This fact, as well as the haste with which he had copied the documents he could find, greatly reduced the value of his publication, which would appear only in 1870, in Florence, as an abstract of the journal *Rivista Europea,* with the title, "The Galileo Trial Reviewed from Newly Found Documents."

After the return of Pius IX to Rome, the fact that the archives of the Holy Office had been violated could not but concern the ecclesiastical authorities. An attempt to fend off possible blows in advance (as in fact the one from Gherardi) was carried out by Marini with his work, *Galileo and the Inquisition: Critical-Historical Papers Addressed to the Roman Academy of Archeology* (1850). Despite the fact that Marini was at that time one of the very few who could freely consult the volume of Galileo's trial (which had, in fact, been placed by Pius IX in the Secret Vatican Archives), he limited himself to sporadic citations. Moreover, Marini's work was predominantly an apology for the behavior of the Holy Office with respect to Galileo and aimed at dissipating the suspicions of an inhumane rigor in the treatment of him during the trial of 1633.

The work by Marini was severely criticized by the Frenchman Henri de l'Epinois, who obtained permission to consult the volume of the trial of Galileo with the intention to publish it. But the copies of the documents were transcribed too hurriedly and were not further checked against the originals. So the work which de l'Epinois published in Paris in 1867, with the title *Galilée, son procès, sa condamnation, d'après des documents inédits,* did not correspond to expectations. Not much better was the work that Domenico Berti published nine years later: *Galileo's Trial Published for the First Time.* On the following year, de l'Epinois published a second, much more accurate edition of the Galilean documents, with the title *Les pieces du procès de Galilée, précédées d'un avant-propos,* which was enriched with interesting historical and explanatory notes. Even better, with still greater critical rigor and scrupulous fidelity to the reproduction of the texts, was the work published that same year (1877) by the German scholar Karl von Gebler, *Proceedings of the Trials of Galileo,* which followed by a year the highly praised 1876 biography of Galileo by the same author: *Galileo Galilei and the Roman Curia.*

Even though the Roman authorities had opened the Secret Vatican Archives, consultations of the archives had been restricted to a very limited number of scholars and under very strict conditions. A more effective liberalization program started only in 1880 when the new Pope Leo XIII opened the archives more broadly. The most conspicuous fruit of it was the complete edition of all of the documents in the archives concerning Galileo's trial, in the nineteenth volume of the national edition of the works of Galileo, edited between 1890 and 1909 by Antonio Favaro.

A little later on, in his encyclical *Providentissimus Deus,* Leo XIII dealt with the problem of the relationship between sacred scripture and science. He based his treatment on theological principles very similar to those used by Galileo in his *Letter to the Grand Duchess Christina.* A reference, at least, to the Galilean problem, which continued to be at the center of disputes in that epoch, profoundly influenced by the spirit of laicism and positivism, would have been more than proper. Instead the pope limited himself to an allusion, formulated in extremely cautious terms, to errors committed by individual Church Fathers and, in following epochs, by their interpreters

> in the case of the explanation of Scriptural passages which deal with physical questions, they held to the opinions of their time, with the result that they perhaps and not always judged truthfully and stated things which are no longer approved today. (Denzinger, *Enchiridion Symbolorum,* no. 1948)

As we see, even in the time of an "open" pope such as Leo XIII, the Church was still quite far from wishing to make even an explicit mention of the Galilean problem. It limited herself to respond in an indirect manner to the attacks and criticisms by encouraging the publication of works of an apologetic character by Catholic authors.

A first attempt on the part of the Catholic Church to overcome this apologetic position occurred with the initiative taken in 1941 by the Pontifical Academy of Sciences for the publication of a biography of Galileo on the occasion, in 1942, of the three-hundredth anniversary of his death. The work was assigned to Monsignor Pio Paschini, professor of Church history at the Pontifical Lateran University in Rome, of which he was also the rector. The president of the academy, the Franciscan Agostino Gemelli, clarified the intent of the planned work by declaring:

> Pio Paschini will give us not just a life but rather he will present us the figure of Galileo by situating his work in the historical framework of the knowledge of his time and by thus putting again the figure of the great astronomer in its true light.

And Gemelli concluded:

> The planned volume will, therefore, be an effective demonstration that the Church did not persecute Galileo but abundantly helped him in his studies. It will not, however, be a work of apologetics, because this is not the task of scientists, but of scientific and historical documentation. (*L'Osservatore Romano,* December 1–2, 1941, 3–4)

The affirmation that the book entrusted to Paschini was not to be an apologetic work was in fact contradicted by the preceding statement that it had to offer "an effective demonstration that the Church did not persecute Galileo." Such a latent contradiction in the working program assigned to Paschini would not fail to become clear immediately after the completion of his work.

Paschini was a well-respected scholar in the field of ecclesiastical history but not an expert in that of Galilean studies. Thus, one cannot help being surprised at his choice, as well as by the implicit wish that he bring to completion such an exacting task in the shortest possible time. In fact, the book, entitled *Vita e Opere di Galileo Galilei* [Life and Works of Galileo Galilei], was completed in only three years and was of considerable length. Even with its clear limitations, due to the lack of the scientific preparation of the author as well as the short time of his contact with the Galilean problem, this work of Paschini contained a rich

documentation and an honest evaluation of the events, not without severe judgments concerning the actions of the Church and of Galileo's
adversaries (in particular of the Jesuit Scheiner). As such it did not please
the chancellor of the Pontifical Academy of Sciences, Pietro Salviucci,
nor Father Gemelli himself. The manuscript was sent, together with a
negative appraisal of it, to the Secretariat of State and then to Pius XII.
The pope knew and esteemed Paschini and transmitted it to the Holy
Office for a second opinion on the matter. The Holy Office, however,
further aggravated the negative judgment, considering the publication
of this work inappropriate. To no avail were the protests of Paschini to
Gemelli and subsequently to then Deputy Secretary of State Montini
(the future Pope Paul VI), who showed him much understanding and
tried to help him. Received later on by the assessor of the Holy Office,
Monsignor Ottaviani, Paschini heard repeated the criticism that he had
done nothing else but written an apology for Galileo, that he had not
given due importance to the fact that Galileo had not brought forth decisive proofs for the Copernican system, and that he — Paschini — had
used expressions and judgments concerning the Church's decisions that
were too extreme. On that occasion Paschini requested and obtained
the restitution of the manuscript, and he kept from then until his death
the strictest silence on what had happened.

Paschini's book, however, came to light two years after the author's
death, in 1964, towards the end of the Second Vatican Council. Its publication coincided with the preparation of the important pastoral constitution *Gaudium et Spes,* whose theme was that of the Church in the
modern world and which was promulgated in December 1965 at the
last session of the council. Given that theme, it was not possible to
omit in it the consideration of the relationship between the vision of
the Christian faith and that of modern science. During the preparatory
phase of the document, the proposal was put forth for a declaration
on Galileo to be inserted in the text. The paragraph prepared as an answer to the proposal said:

> May we be permitted to deplore certain mental attitudes, which are
> alien to healthy scientific research and which in centuries past
> showed themselves visible perhaps within the Church itself. Giv-
> ing birth as they did to disputes and controversies, these mental at-

titudes were the cause whereby many ended by opposing science to faith, with most grave damage to both. On the other end, these errors are easily understood, given those times, and they were not exclusive to Catholics, since similar attitudes were present in other religions [the writer of this paragraph probably wanted to say other religious denominations, such as Protestants]. Still it is necessary that we do our best, in so far as human frailty permits, that such errors, as for example the condemnation of Galileo, are never repeated. (Maccarrone, *Mons. Paschini,* 91)

As we see, even though extremely cautious, it was, nevertheless, an acknowledgment of error. And this did not please the majority of the members of the commission who were responsible for the compilation of the document. The co-president of the commission, Monsignor Pietro Parente, expressed his negative opinion in the following terms:

Galilei. Not appropriate to speak of it in the document, so as not to ask the Church to say it had been wrong. [The Galilean question] should be judged on the basis of those times. In the work of Paschini everything is put in its true light. (Maccarrone, *Mons. Paschini,* 91–92)

As a result, the preference was to insert in the definitive text of *Gaudium et Spes* the following words:

We cannot but deplore certain habits of mind, which are sometimes found too among Christians, which do not sufficiently attend to the rightful independence of science and which, from the arguments and controversies they spark, lead many minds to conclude that faith and science are mutually opposed.

And to the text at this point was added the following note: "See Pio Paschini, *Vita e Opere di Galileo Galilei,* 2 volumes, Vatican Press (1964)."

How was it that Paschini's work, for which permission to publish had been refused twenty years before, and which only two years earlier had still been considered "inappropriate" by Monsignor Parente himself, all of a sudden received the seal of an official document, being even quoted in a council decree? With very few exceptions, such as most probably that of Monsignor Parente, none of the "council fathers" was obviously aware of the way in which such an extraordinary "rehabilitation" of Paschini's work had become possible. In fact, the book, which had by now been published, had been accurately reviewed and conveniently "emended," by toning down, eliminating, or even reversing the author judgments, above all the concluding considerations on the decisions of 1616 and 1633. Surely enough, being published twenty years after its composition, Paschini's text needed to be updated. And, undoubtedly, not a few of the author's interpretations and judgments were debatable. Nobody would therefore have objected to the fact that explanatory notes would have been added to the text, containing the updates and corrections that were deemed necessary. But the most elementary editorial probity would have required that the published text be the original one written by Paschini. It was not so, unfortunately. And it is truly regrettable that a text censured in such a way, and that Paschini would have, no doubt, refused to consider as his own, was published in view of its citation within a solemn declaration of the council on the freedom of scientific research.

Despite these hesitant attitudes and contradictions, which are visible up to the final phases of the Second Vatican Council, the process of rethinking the journey of the Church in the course of history had been by now set forth and could not but go forward. And this made inevitable the explicit consideration, among other subjects, of the Galileo Affair. Pope Paul VI had already made a first, albeit indirect, move in that direction. On the occasion of his speech at the closing of the Eucharistic Congress at Pisa (June 10, 1965), he had said among other things:

> Love, Sons of Tuscany, the Christian faith of this privileged and blessed land, the faith of your Saints, the faith of the great spirits, whose immortal memory has been celebrated yesterday and today, Galileo, Michelangelo and Dante, the faith of your fathers.

It was the first time, since 1633, that a pope mentioned Galileo in a public speech. And now he was included among the "great spirits" whose "immortal memory" and faith were being celebrated. But this "honorable mention" stopped there, without any hint to the facts of the past, and even less to the responsibility of the Church with respect to him.

The first explicit acknowledgment of such responsibility took place only fourteen years later, in 1979. While speaking to the Pontifical Academy of Sciences, on the occasion of the one-hundredth anniversary of the birth of Albert Einstein, the successor of Paul VI, John Paul II, had dedicated ample space even to Galileo, saying, among other things:

> The greatness of Galileo is known to everyone, like that of Einstein; but unlike the latter, whom we are honoring today before the College of Cardinals in the apostolic palace, the former had to suffer a great deal, we cannot conceal the fact, at the hands of men and organs of the Church. The Vatican Council recognized and deplored certain unwarranted interventions.

At this point there followed the citation from *Gaudium et Spes,* which we have reported above. The pope continued:

> To go beyond this stand taken by the Council, I hope that theologians, scholars and historians, animated by a spirit of sincere collaboration, will study the Galileo case more deeply and, in a loyal recognition of wrongs from whatever side they come, will dispel the mistrust that still opposes, in many minds, a fruitful concord between science and faith, between the Church and the world. I give all my support to this task, which will be able to honor the truth of faith and of science and open the door to future collaboration. (John Paul II 1979, n. 6; trans. Bucciarelli, "Speech," 79)

The realization of this proposal was initiated only two year later, with the creation of a commission in charge of the study of the Copernican and Galilean question. In the letter establishing the commission, Secretary of State Cardinal Casaroli expressed the wish that "the work be carried out without delay and that it lead to concrete results."

The commission was constituted with four sections: exegetical, cultural, scientific-epistemological, and historical-juridical. Its work started several months later. Each one of the sections carried on its activities in an autonomous way. The plenary sessions, in which the exchange of ideas and information on the working of the single sections had to take place, were only seven (from October 1981 up to November 1983). After an interval of almost seven years, in May 1990 Secretary Casaroli entrusted to Cardinal Poupard the task of coordinating the final phase of the works of the commission. A little later, Cardinal Poupard requested the coordinators of the single sections to send him a report on the activities achieved and the works published. And on the following July 13 he declared to the members of the commission that the working of the commission had come to an end. There would, however, still be a gap of more than two years before the results obtained were presented to the pope, with a speech of Cardinal Poupard, which was followed by that of the pope himself, on October 31, 1992.

During the time that commission worked, there were published in succession the volume edited by Monsignor (later Cardinal) Poupard, *Galileo Galilei, 350 ans d'histoire 1633–1983,* 1983; *I Documenti del Processo di Galileo Galilei,* edited by Father Sergio Maria Pagano, with the collaboration of A. G. Luciani, 1984; *The Galileo Affair: A Meeting of Faith and Science,* edited by G. V. Coyne, M. Heller and J. Zycinski, 1985; Mario D'Addio, *Considerazioni sui processi di Galileo,* 1985; a series of Galilean studies in English, edited by the Vatican Observatory (1983–1989); Walter Brandmüller, *Galileo e la Chiesa ossia il diritto a errare,* 1992; and *Copernico, Galilei e la Chiesa: Fine della controversia (1820),* edited by Walter Brandmüller and Egon J. Greipl, 1992.

Some of these contributions supply valuable conclusions on particular aspects of the Galileo Affair and its protagonists, or offer important documentary confirmation. Others furnish new documentation of notable value but are vitiated by a lack of a proper scholarly attitude in the evaluation of the documentation, as well as by errors, even serious ones of a historical character. Some others, moreover, are written with an evident apologetic intent that one would have wished to be absent in an "honest research of the responsibilities in the Galileo Affair, from whatever side they come," according to the desire expressed by John Paul II in the speech quoted above. As a matter of fact, the whole project of this Galilean commission was by far beyond the capabilities of the relatively few people to which it had been entrusted, most of whom were ecclesiastics with no previous scholarly competence on the Galileo Affair. This fact, honestly admitted in the important documentary work of Artigas and Sánchez de Toca, will not fail to show its effect in the final report on the commission work presented by Cardinal Poupard, to be discussed below.

In the meantime, John Paul II had repeatedly gone back to treat of Galileo. Speaking before numerous journalists, in May 1983, he declared:

> To you who are preparing to commemorate the 350th anniversary of the publication of the great work of Galileo, *Dialogue Concerning the Two Chief World Systems,* I would like to say that the experience lived by the Church at the time of and following upon the Galileo case, has permitted a maturing and more concrete understanding of the authority which is proper of the Church. . . . Thus is it understood more clearly that divine Revelation, of which the Church is guarantor and witness, does not involve as such any scientific theory of the universe, and the assistance of the Holy Spirit does not in any way come to guarantee explanations that we might wish to maintain on the physical constitution of reality. That the Church was able to go ahead with difficulty in a field so complex should neither surprise nor scandalize. The Church formed by Christ, who has declared himself to be the Way, the Truth, and the Life, remains nonetheless composed of limited human beings who are an integral part of their cultural epoch. (John Paul II, "Discourse to the Symposium," nos. 2 and 3)

Six years later, the Pope spoke again of Galileo, on the occasion of his visit to the city of Pisa and stated:

> How could we not remember at least the name of that great man who was born here and who from here took his first step towards an everlasting fame? Galileo Galilei I speak of, whose scientific work, opposed improvidently in its beginning, is now recognized by all as an essential stage in the methodology of research and, in general, in the path towards the knowledge of the natural world. (John Paul II, "Discourse to the Symposium," no. 2)

John Paul II came back once again to the case of Galileo in a discourse of October 31, 1992. This time it was question not of an occasional mention of Galileo but of a comprehensive judgment, which was intended to be definitive, in this case, as a conclusion of the activities of the commission established thirteen years before. In fact, the papal discourse was preceded by that of Cardinal Poupard, who claimed to be offering, indeed, in a summary fashion, the conclusions of that commission.

The occasion for these discourses was offered by the closure of the plenary session of the Pontifical Academy of Sciences, dedicated to the theme of the emergence of complexity in the field of mathematical, physical, chemical, and biological sciences. But the presence of many cardinals, of a large number of members of the diplomatic corps accredited to the Holy See, and of well-known scientists, members of the Academy, indicated the importance that was attributed to the two discourses, and especially to that of the pope. They would mark the official closure of the Galilean question on the part of the Catholic Church and thus also of a painful chapter of its history, in line with the desire expressed by John Paul II in his discourse of 1979.

Considering the discourse of Cardinal Poupard, one cannot fail to note, first of all, that it offers an evaluation of the positions of two of the protagonists of the Galileo Affair, Cardinal Bellarmine and Galileo himself, which does not agree with the overall judgment derived from the best contributions of the commission. Moreover, it contains an interpretation of the actions taken by the Catholic Church with respect to the Galileo Affair during the eighteenth and nineteenth centuries that is seriously lacking in historical accuracy. And finally, it leaves in the shadows the responsibility "at the top" on the measures taken in 1616 and on the condemnation of Galileo, that is to say, the responsibility of the Holy Office, of the Congregation of the Index, and in the end of the two popes of the epoch, Paul V and Urban VIII.

I begin with the presentation of Bellarmine's position. According to Cardinal Poupard, in the *Letter to Foscarini* of April 12, 1615, Bellarmine had correctly formulated the problems posed by the system of Copernicus. Are there real and verifiable proofs for it? Is it compatible with scripture? According to Robert Bellarmine, continued the cardinal:

As long as there was no proof that the Earth orbited around the Sun, it was necessary to interpret with great circumspection the biblical passages declaring the Earth to be immobile. If the orbiting of the Earth were ever demonstrated to be certain, then the theologians, according to him, would have to review their interpretations of the biblical passages apparently opposed to the new Copernican theories, so as to avoid asserting the error of opinions which had been proved to be true. (Poupard, *Toward a Resolution,* no. 2)

At this point, Cardinal Poupard quoted Bellarmine:

I say that if it were really demonstrated that the Sun is at the center of the world and the Earth is in the third heaven, and that it is not the Sun which revolves around the Earth, but the Earth round the Sun, then it would be necessary to proceed with great circumspection in the explanation of the Scriptural texts which seem contrary to this assertion and to say that we do not understand them, rather than to say that what is demonstrated is false. (Ibid.)

There is an obvious contrast between Cardinal Poupard's summary of Bellarmine's position and the words of the latter that he quotes. As it is evident from them, Bellarmine speaks of the circumspection on the part of the theologians not as long as there were no proofs of the Earth's rotation around the Sun, but in the case that such proofs existed. And, paradoxically, Bellarmine seems to imply that circumspection is not at all necessary in the absence of such proofs.

The inaccuracy of Poupard's summary, which tends to present Bellarmine's position as one of detached objectivity, becomes all the more evident if one looks at those phrases of the Jesuit cardinal in the whole context of his answer to Foscarini. On this matter, note what I have already remarked in chapter 3 of this book. I will limit myself here to stress once more the fact that the apparent admission, on the part of Bellarmine, of the possibility of future proofs of Copernicanism is in practice denied both by the theological motifs given in the first part of the letter and by the exegetical and philosophical ones adduced in the final part of it.

On the other hand, it seems to me important to note that if Bellarmine had really been convinced of the possibility of a future proof of the Copernican theory, he should have suggested to Paul V not to take a precipitous decision that would have excluded that possibility. But, as is well known, he was on the contrary in agreement with the decisions taken by the pope at the end of February 1616 and was even given the task of informing Galileo privately that, since it was contrary to scripture, the Copernican theory could not be held.

As to Galileo's position, Cardinal Poupard comments:

> In fact, Galileo had not succeeded in proving irrefutably the double motion of the Earth, its annual motion around the Sun and its daily rotation on the polar axis, when he was convinced that he had found proof of it in the ocean tides, the true origin of which only Newton would later demonstrate. Galileo proposed a tentative proof in the existence of the trade winds, but at that time no one had the knowledge for drawing therefrom the necessary clarifications. More than a hundred and fifty years still had to pass before the optical and mechanical proofs for the motion of the Earth were discovered. (Ibid., no. 3)

Even those statements truly leave one perplexed. In the first place, I think it is necessary to point out that the scientific value of the arguments proposed by Galileo for the Copernican system were not considered at all, neither in 1616 nor at the time of Galileo's trial. In 1616, the so-called "proof of the tides" was not made public by Galileo, but only privately consigned to Cardinal Orsini in manuscript form, in hopes of garnering from him a favorable intervention with Paul V. It was not submitted for examination to the experts of the Holy Office and did not have, as a consequence, any influence on the decisions of the latter, which brought forth the decree of the Congregation of the Index against the Copernican system. But even if the qualifiers-consultors of the Holy Office could have examined this proof from the tides, they would have excluded a priori its validity, not because of "scientific" reasons, but only because the Copernican hypothesis, which that proof intended to support, was "foolish and absurd from the point of view of philosophy," that is, of the natural philosophy of Aristotle, the only one believed by them to be true. As to 1633, no document at our disposal shows that that proof, expressed in the *Dialogue,* was even then taken into consideration.

There remains, however, the fact, which Cardinal Poupard omits to consider, of the discoveries made by Galileo with the telescope. They were sufficient to prove that fundamental affirmations of the Aristotelian-Ptolemaic theory were untenable, as the Jesuit mathematician Clavius himself had acknowledged. Surely enough, there remained still the possibility of the geocentric system of Tycho Brahe. But beside it there existed at least an equal possibility for the Copernican system. Now, in 1616 the qualifiers-consultors of the Holy Office did not take at all into consideration this fact, which should have induced Paul V and the cardinals of the Holy Office not to precipitate a decision that could prove itself to be wrong, as in fact was the case.

As to interpreting the history of the measures taken by the Church during the eighteenth and nineteenth centuries, Cardinal Poupard states:

> For their part, Galileo's adversaries, neither before nor after him, have not discovered anything which could constitute a convincing refutation of Copernican astronomy. The facts were unavoidably clear, and they soon showed the relative character of the sentence passed in 1633. This sentence was not irreformable. In 1741, in the face of the optical proof of the fact that the Earth revolves round the Sun, Benedict XIV had the Holy Office grant the *imprimatur* to the first edition of the *Complete Works of Galileo.* This implicit reform of the 1633 sentence became explicit in the decree of the Sacred Congregation of the Index which removed from the 1757 [*sic*] edition of the Catalogue of the Forbidden Books works favoring the heliocentric vision. (Ibid., nos. 3 and 4)

It seems to me necessary to comment, in the first place, that the *imprimatur* for the edition at Padua of the works of Galileo was granted on the condition that the sentence of condemnation and the abjuration of Galileo be reproduced at the beginning of the volume containing the *Dialogue.* The editors eliminated Galileo's marginal notes, so that the *Dialogue* could be presented as a pure mathematical hypothesis. Also, the edition was not, in fact, complete, since the *Letter to the Grand Duchess Christina* was excluded. It is not possible, therefore, to see in this Paduan edition of Galileo's works an "implicit reform" of the 1633 sentence. Nor is it possible to agree with what Cardinal Poupard claims immediately thereafter, namely, that the implicit reform "became explicit in the decree of the Sacred Congregation of the Index of 1757." That decree had nothing to do with the sentence of 1633. It only decided to omit in the new edition of the Index (which appeared in 1758) the general prohibition of books on Copernicanism, which had not been specifically condemned by name. But it left on the Index, among those which had been specifically condemned, the works of Copernicus, Galileo, and Kepler.

On the other hand, the ambiguity of the 1757 decision is proven by the well-known Settele affair of 1820. After having recalled the refusal of the Master of the Sacred Palace Anfossi to give an *imprimatur* to the works of the Canon Settele, *Elements of Optics and Astronomy* (in fact it was only a question of the second volume, *Elements of Astronomy*), Cardinal Poupard states: "The incident gave the impression that the 1633 sentence had indeed remained unreformed because it was irreformable." The sentence in favor of Settele, Cardinal Poupard recalls, was obtained thanks to the favorable report prepared by the commissary of the Holy Office, Olivieri. But that report, as I have already discussed, contained a completely absurd interpretation of the decree of 1616 and of Galileo's condemnation, which excluded the necessity of their "reform." As we have already seen, Olivieri had claimed that the Church had condemned Copernicanism in 1616 and the *Dialogue* in 1633 not because of their opposition to scripture but for the deficiency of the heliocentric hypothesis in the light of the scientific knowledge of the epoch. The Church could now allow, without contradiction, the adoption of the Copernican system "as it was by now commonly seen," that is, corrected thanks to the discovery by Torricelli of the gravity of air and justified by Newtonian dynamics and subsequent astronomical discoveries. On the other hand, the solution of the Settele affair in 1820–22 did not imply in any way the same for the Galileo Affair, as is proven by the fact that the *Dialogue,* as well as the *On the Revolutions* of

Copernicus and the *Epitome* of Kepler were eliminated from the Index only in the edition of 1835.

At the conclusion of his discourse Cardinal Poupard affirms that the philosophical and theological qualifications that were "wrongly" given to the theory of the centrality of the Sun and mobility of the Earth were the result

> of a transitional situation in the field of astronomical knowledge, and of an exegetical confusion regarding cosmology. It is in that historical and cultural framework, far removed from our own times, that Galileo's judges, incapable of dissociating faith from an age-old cosmology, believed, quite wrongly, that the adoption of the Copernican revolution, in fact not yet definitively proven, was such as to undermine Catholic tradition, and that it was their duty to forbid its being taught. This subjective error of judgment, so clear to us today, led them to a disciplinary measure from which Galileo had much to suffer. These mistakes must be frankly recognized as you, Holy Father, have requested. (Poupard, *Toward a Resolution,* no. 5)

As we see in these concluding statements of the cardinal's discourse, the condemnation of Galileo is presented as a "disciplinary measure." No mention is made of the Church's decisions in 1616 that constitute the factual basis, both theological and juridical, for that condemnation. Now, the decree of the Index of March 1616 was certainly not a "disciplinary measure," but a decree that reported the doctrinal decision made by Paul V while declaring that the Copernican doctrine, as an explanation of the real structure of the world, was "false and totally contrary to Divine Scripture" (Galileo, *Opere,* 19:323). It was the intention of Paul V and of the cardinals of the Inquisition and those of the Congregation of the Index that that decree was definitive and, therefore, certainly not seen by them as relative and reformable. Bellarmine's admonition and the subsequent injunction of Commissary Segizzi had, on the other hand, a juridical character. Such juridical character is certainly present in the 1633 trial. But on that occasion Urban VIII, as we already know, determined on the doctrinal plane the theological extent of the opposition of Copernicanism to scripture, which had been left unclear by the 1616 decree, defining it as an error in faith, thus justifying Galileo's condemnation for a "vehement suspicion of heresy." In the end, this attempt to reduce the sentence of 1633 to a simple "disciplinary measure," taken by quite unspecified "judges of Galileo," without any mention of those responsible "at the top," the cardinals of the Holy Office and Urban VIII himself, seems to be an attempt to minimize Galileo's condemnation and his abjuration, even if the words of John Paul II in his discourse of 1979 are added: "from which Galileo had much to suffer."

That same tendency to scale down the Church's responsibility in the Galileo Affair, by emphasizing that of Galileo, is evident in the papal discourse. Right from the beginning of it, John Paul II stresses the existence of a double paradox. On one hand Galileo, "rejecting the suggestion to present the Copernican system as an hypothesis, inasmuch as it had not been confirmed by irrefutable proofs," did not abide by the exigencies of that same experimental method "of which he was the inspired founder." On the other hand, most of the theologians erred on the plane of biblical exegesis, of their own competence, to which, on the contrary, Galileo, a simple layman, proved to possess a better understanding.

The pope then treats of the pastoral concern that influenced the Church's judgment on the question of Copernicanism at the time of Galileo. "By virtue of her own mission," he states,

> the Church has the duty to be attentive to the pastoral consequences of her teaching. Now, the pastoral judgment which the Copernican theory required was difficult to make, insofar as geocentrism seemed to be a part of scriptural teaching itself. Undoubtedly, the pastor ought to show a genuine boldness but he must avoid the double trap of a hesitant attitude and of hasty judgment, both of which can cause considerable harm. (John Paul II, "Discourse on the Occasion," no. 9)

At this point John Paul II recalls a crisis analogous to that of Copernicanism, the one concerning biblical studies at the end of the nineteenth and the beginning of the twentieth centuries. Facing the dangers to the faith from the rationalistic tendencies in scriptural studies prevalent at the time, "certain people thought it necessary to reject firmly based historical conclusions. That was a hasty and unhappy decision."

One of the causes of Galileo's condemnation, the pope continues, was the fact that:

> The vast majority of theologians did not recognize the formal distinction between Sacred Scripture and its interpretation, and this led them unduly to transpose into the realm of the doctrine of faith a question which, in fact, pertained to scientific investigation. (Ibid.)

And here, in agreement with what Cardinal Poupard had stated, the pope points out the exception of Bellarmine, "who had seen what was truly at stake in the debate." As a confirmation of this claim John Paul II quotes the well-known words of the Jesuit cardinal in his *Letter to Foscarini* on the great circumspection to be used in explaining Scripture, should there be eventual scientific proofs in favor of the motion of the Earth, "and say that we do not understand, rather than to affirm that what has been demonstrated is false."

The pope also takes up the statement of Cardinal Poupard "that the sentence of 1633 was not irreformable, and that the debate, which had not ceased to evolve, thereafter, was closed in 1820 with the *imprimatur* given to the work of Canon Settele." And he adds:

> From the beginning of the Age of Enlightenment down to our own day, the Galileo case has been a sort of "myth," in which the image fabricated out of the events was quite far removed from reality. In this perspective, the Galileo case was the symbol of the Church's supposed rejection of scientific progress, or of "dogmatic" obscurantism opposed to the free search of truth. This myth has played a considerable cultural role. It has helped to anchor a number of scientists of good faith in the idea that there was an incompatibility between the spirit of science and its rules of research on the one hand and the Christian faith on the other. A tragic mutual incomprehension has been interpreted as the reflection of a fundamental opposition between science and faith. The clarifications furnished by

recent historical studies enable us to state that this sad misunderstanding now belongs to the past. From the Galileo Affair we can learn a lesson which remains valid in relation to similar situations which occur today and may occur in the future. (Ibid., nos. 9 and 10)

Several statements of this discourse, as of that of Cardinal Poupard, which has furnished its starting point, cannot fail to perplex Galilean scholars. Such is, in the first place, the criticism of Galileo for having refused to accept the Copernican theory as a mere hypothesis, in the absence of certain proofs, with the result that he did not abide by the very exigencies of the experimental method. However, the term "hypothesis," here used without further specification by the pope, is completely ambiguous. Setting aside the meaning of this term in modern science, for Galileo it meant a proposal or an attempt to explain the nature and the causes of natural phenomena whose explanation was not yet corroborated by "sense experience" and by "certain demonstrations, but could be so corroborated in the future." Galileo was well aware that Copernicanism, without such experiences and demonstrations, remained a simple hypothesis in that sense of the word. In agreement with many Galileo scholars, I do not think that he ever claimed to have certain proofs of Copernicanism. But even if one wished to claim that for him the argument from the tides was a "certain demonstration," that would not justify blaming him for not having respected the requirements of experimental method. If anything, the accusation would be that he had erred by considering as valid an argument that was not so. But this is an error of judgment (found not infrequently even in the greatest of modern scientists), not an indication of infidelity to the experimental method.

What Galileo refused to accept was that the Copernican system was a hypothesis in the sense of a purely mathematical model, as was stated by Osiander in the prefatory "Notice to the Reader," added to the beginning of *On the Revolutions*. We already know that this was the traditional view of astronomical theories in antiquity and in the Christian Middle Ages. And Cardinal Bellarmine, in the first part of his *Letter to Foscarini*, made, in fact, reference to the merely mathematical meaning when he suggested in an indirect way to Galileo that he stick to it. A hypothesis taken in this sense had nothing to do with experimental method. It does not seem, therefore, that Galileo can be accused of not having abided by the requirements of such a method when he refused to consider the Copernican system as a hypothesis of this kind.

Nor can we agree with the other affirmation of the papal discourse, according to which, in his response to Foscarini, the Jesuit cardinal, in contrast with the majority of the theologians of his time, had perceived what was "truly at stake in the debate." This is a statement in strict agreement with the one I have already considered in the discourse of Cardinal Poupard and which, therefore, lends itself to the same critical remarks.

As to the statement that the 1633 sentence "was not irreformable," one must once more remark that at the time of Galileo's condemnation neither the cardinals of the Holy Office nor Urban VIII himself certainly thought it to be "reformable." How to justify, otherwise, the severity of the condemnation, including the abjuration, and of the subsequent treatment of Galileo up to his death and even afterward? Why not admit it frankly, instead of trying to diminish the gravity of such decisions? On the other hand, if that sentence was not "irreformable" the problem remains of why it has never been officially reformed, and likewise for the decree of the Index of 1616, which constituted the starting point for the juridical justification of Galileo's condemnation. It does not seem to me that one could affirm with regard to this, as it is stated both in the discourse of Cardinal Poupard and in the papal one, that the decisions taken in 1820–22 with regard to the Settele affair represented the "closure" of the Galileo case. As already noted, such decisions did not concern Galileo at all, nor the decree of 1616, but only the teaching of the Copernican system. And according to the "remarkable" way out of the centuries-old impasse excogitated by the commissary of the Holy Office in 1820, the Church had been right in rejecting the original Copernican system because of its deficiencies in light of the scientific knowledge of the epoch. For the same reasons the sentence of condemnation of the *Dialogue* and of Galileo himself was thus also justified. An extraordinary way, indeed, to close the Galilean debate!

Furthermore, both Cardinal Poupard's discourse and that of the pope give the impression of a noticeable change of position with respect to the initial papal discourse of 1979. While in the latter there had been a very short but nevertheless explicit acknowledgment that

Galileo had suffered "a great deal at the hands of men and organs of the Church," in these last two discourses the ecclesiastical responsibility is attributed to a group of theologians. (Poupard speaks also of "judges," but even this term does not appear in the papal discourse.) No mention is made of the organs of the Church (the Holy Office and the Congregation of the Index), nor of those at the top responsible for Galileo's condemnation, Paul V and Urban VIII.

On the other hand, Galileo is now implicitly represented as equally responsible for the "tragic misunderstanding" that brought about his condemnation. How could we, in fact, interpret differently the pope's words of a "tragic mutual incomprehension," which seem to constitute the final judgment on the Galilean question? But can we, in fact, consider Galileo to be responsible for his part in that "tragic incomprehension"? One can surely consider him to be responsible of tactical mistakes in his activities in support of the Copernican theory, but not of a tragic incomprehension of the theoretical aspects of the issue, neither on the scientific issues nor on the theological ones. Bellarmine, on the contrary, in both discourses, is depicted as one of the few "illuminated" theologians, capable of giving a lesson of scientific methodology to Galileo as well as one of biblical exegesis to his colleagues, the theologians.

As to Galileo's condemnation, Cardinal Poupard had limited himself to mention it as a mere "disciplinary measure." And John Paul II, as I have already noted, in mentioning Poupard's affirmation, only repeats that that condemnation was not irreformable. Too little indeed, in a discourse that wanted to mark the definitive closure, on the part of the Church, of the Galileo Affair!

In conclusion, the discourse of Cardinal Poupard and even more so the papal discourse because of its greater authority, cannot but leave disappointed those who expected that the official conclusion of the Galileo commission would be in line with the desire expressed by John Paul II in 1979 that there would be a "loyal recognition of wrongs from whatever side they come." The disappointment is increased when one realizes that the majority of the studies published under the auspices of the commission appear to be in line with the pope's desires and are open to a frank admission of errors, even at the highest level of ecclesiastical hierarchy. And so one cannot help but ask why, at the moment of the Church's final and authoritative conclusion, so much weight had been given to judgments in some of the commission's contributions, which tended to emphasize the responsibility of Galileo in such a way as, in the end, to minimize the gravity of his condemnation by presenting it as a simple disciplinary (and reformable) measure.

All that I have pointed out should not be taken as a personal criticism of John Paul II. His intentions were, no doubt, sincere and honest. That the final result of his proposal did not correspond to what he had wished for in his discourse of 1979, did not depend, I think, upon him, but rather on those who reported to him on the results of the commission's work, and who collaborated in the drafting of his "closing" discourse (as is known, the papal discourses, apart from rare exceptions, are prepared by "experts"). They appear to have been once more strongly influenced in their judgments by the concern, which had already prevailed in their predecessors of the past, that is, that of closing, indeed, the Galileo Affair, but at the same time and above all, of saving the decorum of the Church.

Epilogue

This book started with a prologue. I have purposefully used such a term, which recalls the idea of an introductory discourse to a drama or to an ancient tragedy. The Galileo Affair is, indeed, a drama and from many respects even a tragedy. A drama and a tragedy more than four hundred years old by now, but whose historical existence spans from then to the present day. At the end of these pages, whose intent was, in fact, to narrate this drama, I feel thus authorized to continue using such a dramatic terminology and give the title of "epilogue" to my final considerations.

An epilogue is in fact the conclusion of a drama or of a tragedy. But in which sense one can speak of conclusion in the case of the Galilean drama? Several comments of the Italian and foreign press have tried to see such a conclusion in the discourses of Cardinal Poupard and especially of John Paul II, which I have examined at the end of the last chapter. As we have seen, at the news of Galileo's death, his old "friend" Urban VIII had once again accused him of having held "a very false and very erroneous opinion" and of having given "much scandal to the universal Church." Exactly 350 years later, Urban's successor has acknowledged the mistake of Galileo's condemnation and the role that his drama has played in a "more correct understanding of the authority which is proper to the Church" and its function in "teaching" the Church. With all the reservations that I have made concerning these discourses, one cannot but take note that a long and painful chapter of the Church's history has thus somehow been brought to a conclusion.

This, however, does not mean in any way that the Galileo case has been closed. It is not, and it cannot be so, first of all, from the view point of historical research, since the latter, by its own nature, is essentially an open project. And the contributions of the commission for the study of the Galilean question certainly cannot pretend to have closed it. The opening to scholars of the archives of the former Holy Office, which took place in 1998, is already offering a new and rich documentation on the historical and juridical aspects of the Church's activity at the time of Galileo and in the period subsequent to his condemnation. Such a picture is essential for a better understanding of the events of Galileo's life and, more in general, of those concerning the acceptance of the Copernican vision of the world.

Neither can the Galileo Affair be considered closed by virtue of a pretended "rehabilitation" of the scientist, which would have been achieved, in fact, (again, according to some press comments) by the papal discourse of 1992. In juridical terminology, rehabilitation means the reintegration of the rights and of the reputation of a person who has been the object of a condemnation. Now, it is certainly true that the 1633 condemnation implied for Galileo the loss of many and important rights and, in the intentions of Urban VIII and of the Holy Office, also of his reputation, through the distribution (altogether exceptional in its amplitude) of the sentence of his condemnation in the Italian states of the epoch and in Catholic Europe. In spite, however, of ecclesiastical intentions, Galileo never lost his reputation before the more enlightened people of those times. And the rightness of his position, as well as the injustice of his condemnation, were acknowledged in the following centuries in an ever larger strata of the cultural world, even of the Catholic one. In 1992, surely, Galileo did not need any rehabilitation. And one has to recognize that the intention of such a rehabilitation had been officially excluded by the Vatican itself at the time of the institution of the Galilean commission.

There is, however, a further and deeper sense in which the Galileo case is not, and cannot, be considered as closed. The papal discourse of

1992 affirmed that the painful misunderstanding that has given birth to the myth of an unavoidable opposition of the Catholic Church to scientific progress "belongs by now to the past." In reality, the dialogue of the Church with science remains even today complex and difficult. The more so if one takes into account the width and gravity of the problems that contemporary science and technology present to the reflection of faith. Let us only consider, in the field of scientific research, the problem of evolution and, in that of technology, the problems concerning the development of the new biogenetic technologies, as well as those of birth control. In such a situation, the confrontation between faith and science may become even more traumatic than at the time of Galileo.

Does this mean that a new Galileo case is possible today? If one considers only the juridical aspects of Galileo's condemnation, the answer is certainly negative. The Holy Office does not exist anymore. Even if its successor, the Congregation for the Doctrine of Faith, remains in charge of doctrinal control, its effective power is limited (at least directly and openly) to the sphere of secular clergy and religious orders and congregations. No scientist could be obliged, today, to the abjuration before an ecclesiastical tribunal of his scientific opinions and even less be confined to house arrest. Even the possibility of using the "secular arm," that is to say, the Catholic political parties, in order to extend the doctrinal and disciplinary authority of the Church well beyond the boundaries of the ecclesiastical world, has become more and more problematic. A right sense of laicism, quite different from the old anticlericalism, makes such maneuvering less and less feasible and acceptable, even in Catholic countries.

Undoubtedly, science and technology cannot pretend to solve by themselves alone the problems that their prodigious development lays before all of humanity. It is, however, equally true that the Catholic Church cannot pretend to solve them with doctrinal declarations on a dogmatic or biblical basis. If Galileo's case has taught us something, it is, in fact, to show the danger of similar hasty declarations, which run the risk of becoming untenable in the future and which, on the other hand, the contemporary lay Catholic world is less and less disposed to accept because less and less able to understand them. It is sufficient to think, in this context, of the present status in the Catholic world of the prohibition of all artificial birth control techniques, sanctioned eighty years ago by Pope Pius XI, with his encyclical letter *Casti connubii*. Based as it was on a very disputable biblical foundation, as well as on the traditional Thomist philosophical view of marriage, it could be considered as a sort of new Galileo case, with far-reaching consequences in the everyday life of millions of people. As it happened with the progressive disregard of the prohibition of the Copernican theory, in the post-Galileo era, one can see the large majority of those who still consider themselves faithful Catholics using a wide range of birth control techniques, with the Church officially still defending the status quo.

In reality, the dialogue with science cannot be an exclusive prerogative of the Catholic Church. Other great Christian religious denominations also exist. An authentic ecumenical spirit cannot refuse to take into account the existence of reflection that also takes place among them concerning the problems that contemporary science and technology pose to the common Christian patrimony of faith. Neither does it seem possible to ignore the great non-Christian religions and their contributions to the dialogue between science and religious faith. In fact, such a dialogue cannot be limited to the Christian cultural world alone. This is particularly true in the case of Judaism and Islamism, because of the common monotheistic faith based on the teaching of a holy book—the Bible, as in Christianity, for the former, and the Qur'an for the latter—considered as the revealed word of God. As a result, the relationship between faith and science (and technology) presents to these religions challenges very similar to those in the Christian world. In conclusion, such a dialogue cannot be limited to the Christian cultural world alone.

To go even further, one must take note of the fact that today's science does not have, as its sole counterpart, religious faith. Beside the latter, and more and more independently from it, there exists contemporary thought, which includes philosophy in all its branches, as its basis, and moreover historical, social, political and economic sciences. In Galileo's time, as we have seen, philosophy was still deeply connected with theology, constructing with it the Christian vision of the world. Such a too strict connection was, indeed, at the origin of the Galileo Affair. Galileo as a philosopher saw already the necessity of a new philosophy, certainly not in opposition, according to him, to theology, but nevertheless independent of it. Over a long period of time this new philosophy has become more and more secular, deeply influencing the sociopolitical, economic, and ethical perspectives of contemporary society. To be sure, it is a pluralistic way of thinking, often with contradictory tendencies. But it is impossible to ignore its existence and underestimate its importance, even concerning the problems brought up by contemporary science and technology.

In conclusion, the teaching role for the future, which John Paul II himself has attributed to the Galileo case, demands on the part of the Catholic Church readiness to carry on an attentive, patient, and humble dialogue with the whole of contemporary religious and secular thought. Only through such a dialogue will emerge once more a common project for the solution of problems enormously more complex than the one exhibited four hundred years ago by the Copernican question. Our remaining hope is that Galileo will not have suffered in vain.

Applebaum, Wilbur, and Renzo Baldasso. "Galileo and Kepler on the Sun as Planetary Mover." In Montesinos and Solís, *Largo campo di filosofare*, 381–90.

Artigas, Mariano. "Un nuovo documento sul caso Galileo: EF 2291." *Acta Philosophica* 10, fasc. 2 (2001): 199–214.

Artigas, Mariano, and Melchior Sanchez de Toca. *Galileo e il Vaticano*. Venice: Marcianum Press, 2009.

Baldini, Ugo. *Legem impone subactis: Studi su filosofia e scienza dei gesuiti in Italia 1540–1632*. Rome: Bulzoni Editore, 1992.

Baldini, Ugo, and George V. Coyne. *The Louvain Lectures of Bellarmine and the Autograph Copy of His 1616 Declaration to Galileo*. Vatican City State: Vatican Observatory Publications, 1984.

Beltrán Marí, Antonio. *Talendo y Poder*. Pamplona: Laetoli, 2006.

Beretta, Francesco. *Galilée devant le Tribunal de l'Inquisition*. Fribourg: 1998.

———. "Le Procès de Galilée et les Archives du Saint-Office: Aspects judiciaires et théologiques d'une condamnation célèbre." *Revue des Sciences Philosophiques et Théologiques* 83, no. 3 (July 1999): 441–90.

———. "Le Siège Apostolique et l'Affaire Galilée: Relectures romaines d'une condamnation célèbre." *Roma moderna e contemporanea* 7, no. 3 (Sept–Dec, 1999): 423–61.

———. "L'Archivio della Congregazione del Sant'Ufficio, L'Inquisizione Romana." In *L'Inquisizione Romana: Metodologia delle Fonti e Storia Internazionale; Atti del Seminario Internazionale Montereale Valtellina, 23–24 Settembre 1999*, 119–45. Trieste: University of Trieste, 2000.

———. "Un nuovo documento sul processo di Galileo Galilei: La lettera di Vincenzo Maculano del 22 aprile 1663 al cardinale Francesco Barberini." *Nuncius* 16, no. 2 (2001): 629–41.

————. "Omnibus Christianae, Catholicaeque Philosophiae amantibus, D. D. Le Tractatus syllepticus de Melchior Inchofer, censeur de Galilée." *Freiburger Zeitshrift für Philosophie und Theologie* 48 (2001): 301–25.

————. "Urban VIII Barberini protagoniste de la condamnation de Galilée." In Montesinos and Solís, *Largo campo di filosofare,* 549–74.

Biagioli, Mario. *Galileo Courtier.* Chicago: University of Chicago Press, 1993.

Blackwell, Richard J. *Galileo, Bellarmine, and the Bible.* Notre Dame, IN: University of Notre Dame Press, 1991.

————. *Behind the Scenes at Galileo's trial.* Notre Dame, IN: University of Notre Dame Press, 2006.

Brandmüller, Walter. *Galileo e la Chiesa ossia il diritto ad errare.* Vatican City State: Libreria Editrice Vaticana, 1992.

Brandmüller, Walter, and Egon J. Greipl. *Copernico, Galilei e la Chiesa.* Florence: Leo S. Olschki, 1992.

Brecht, Bertolt. *Galileo.* Translated by Charles Laughton. New York: Grove Press, 1952.

Brodrick, James. *Robert Bellarmine, Saint and Scholar.* Westminster: Newman Press, 1961.

Bucciantini, Massimo. "Dopo il Sidereus Nuncius: Il copernicanesimo in Italia tra Galileo e Keplero." *Nuncius* 9, no. 1 (1994): 15–35.

————. *Contro Galileo: Alle origini dell'Affaire.* Florence: Leo S. Olschki, 1995.

————. *Galileo e Keplero.* Torino: Giulio Einaudi, 2003.

Bucciarelli, Brenno. "Speech of His Holiness John Paul II on Einstein—Galileo." In *Einstein, Galileo: Commemoration d'Albert Einstein.* Vatican City State: Libreria Editrice Vaticana, 1980, 77–81.

Burstyn, Harold L. "Galileo's Attempt to Prove that the Earth Moves." *Isis* 53 (1962): 161–85.

Camerota, Michele. *Galileo Galilei e la cultura scientifica dell'età della Controriforma.* Rome: Salerno Editrice, 2004.

Campanella, Tommaso. *Apologia per Galileo.* Edited by S. Femiano. Milan: Marzorati, 1971.

————. *Apologia pro Galileo.* Translated by M.-P. Lerner. Paris: Les Belles Lettres, 2001.

Carugo, Adriano, and Alistair C. Crombie. "The Jesuits and Galileo's Ideas of Science and of Nature." *Annali dell'Istituto e Museo di Storia della Scienza di Firenze* 8, no. 2 (1983): 3–46.

Cerbu, Thomas. "Melchior Inchofer, un homme fin et rusé." In Montesinos and Solís, *Largo campo di filosofare,* 587–611.

Cioni, Michele. *I documenti Galileiani del S. Uffizio di Firenze.* Florence: Libreria Editrice Fiorentina, 1908.

Clagett, Marshall. *Greek Science in Antiquity.* New York: Abelard-Schuman, 1971.

Clavelin, Maurice. *The Natural Philosophy of Galileo.* Translated by A. J. Pomerans. Cambridge, MA: MIT Press, 1974.

———. "Le 'Dialogue ou la conversion rationnelle': À propos de la première journée." In Galluzzi, *Novità celesti e crisi del sapere,* 17–29.

———. "Galilée, astronome philosophe." In Montesinos and Solís, *Largo campo di filosofare,* 19–39.

———. *Galilée copernicien: Le premier combat, 1610–1616.* Paris: A. Michel, 2004.

Coyne, George V., and Ugo Baldini. "The Young Bellarmine's Thoughts on World Systems." In Coyne, Heller, and Zycinski, 103–10.

Coyne, George V., Michal Heller, and Jozef Zycinski, eds. *The Galileo Affair: A Meeting of Faith and Science; Proceedings of the Cracow Conference.* Vatican City State: Vatican Observatory Publications, 1985.

Crombie, Alistair C. *Augustine to Galileo.* 2 vols. London: Falcon Press, 1969.

D'Addio, Mario. *Il caso Galilei: Processo, Scienza, Verità.* Rome: Studium, 1993.

Dame, Bernard. "Galilée et les taches solaires." In *Galilée: Aspects de sa vie et de son oeuvre,* ed. S. Delorme, 186–251. Paris: Presses Universitaires de France, 1968.

D'Elia, Pasquale M. *Galileo in China: Relations through the Roman College between Galileo and the Jesuit Scientist-Missionaries (1610–1640).* Translated by M. Sciascia. Cambridge, MA: Harvard University Press, 1960.

Denzinger, Heinrich. *Enchiridion Symbolorum.* Edited by K. Rahner. Herder: Rome, 1967.

Dorn, Matthias. *Das Problem der Autonomie der Naturwissenshaften bei Galilei.* Stuttgart: Steiner, 2000.

Drake, Stillman. *Discoveries and Opinions of Galileo.* New York: Doubleday, 1957.

———. *Dialogue Concerning the Two Chief World Systems.* Berkeley: University of California Press, 1967.

———. *Galileo Studies.* Ann Arbor: University of Michigan Press, 1970.

———. *Galileo at Work: His Scientific Biography.* Chicago: University of Chicago Press, 1978.

———. *Galileo.* New York: Hill and Wang, 1980.

———. *Telescopes, Tides, and Tactics.* Chicago: University of Chicago Press, 1983.

———. "Reexamining Galileo's Dialogue." In *Reinterpreting Galileo,* ed. W. A. Wallace, 155–75. Washington DC: Catholic University of America Press, 1986.

———. "Galileo's Steps to Full Copernicanism and Back." *Studies in History and Philosophy of Science* 18 (1987): 93–105.

———. *Galileo: Pioneer Scientist.* Toronto: University of Toronto Press, 1990.

—. *Essays on Galileo and the History and Philosophy of Science.* Edited by Noel Swerdlow and Trevor Levere. 3 vols. Toronto: University of Toronto Press, 1999.

Dreyer, Johann L. E. *A History of Astronomy from Thales to Kepler.* Revised edition, edited by W. H. Stahl. New York: Dover Publications, 1953.

Duhem, Pierre. *Essai sur la notion de théorie physique de Platon à Galilée.* Paris: Hermann, 1908.

—. *Le Système du Monde.* 10 vols. Paris: Hermann, 1913–59.

Fabris, Rinaldo. *Galileo Galilei e gli orientamenti esegetici del suo tempo.* Rome: Pontificia Academia Scientiarum, 1986.

Fantoli, Annibale. *Galileo and the Catholic Church: A Critique of the "Closure" of the Galileo Commission's Work.* Translated by George V. Coyne. Studi Galileiani 4.1. Vatican City State: Vatican Observaory, 2002.

—. *Galileo: For Copernicanism and for the Church.* 3rd ed. Vatican City State: Libreria Editrice Vaticana, 2003.

—. "The Disputed Injunction and Its Role in Galileo's Trial." In *The Church and Galileo,* ed. Ernan McMullin, 117–49. Notre Dame, IN: University of Notre Dame Press, 2005.

—. *Extraterrestri: Storia di un'idea dalla Grecia a oggi.* Rome: Carocci, 2008.

—. *Galileo: Per il Copernicanesimo e per la Chiesa.* 3rd ed. Vatican City State: Libreria Editrice Vaticana, 2010

—. *Galileo e la Chiesa: Una controversia ancora aperta.* Rome: Carocci, 2010.

Favaro, Antonio. *Carteggio inedito di Ticone Brahe, Giovanni Keplero e di altri celebri astronomi e matematici dei secoli XVI e XVII con Giovanni Antonio Magini.* Bologna: Zanichelli, 1886.

—. "La Libreria di Galileo Galilei descritta ed illustrata." *Bolletino di Bibliografia e Storia delle Scienze Matematiche e Fisiche* 19 (1886): 219–93.

—. *Miscellanea Galileiana Inedita.* Venice: Carlo Ferrari, 1887.

—. *Bibliografia Galileiana: Indici e cataloghi.* Vol. 16 Rome-Florence: Fratelli Bencini, 1896.

—. *Oppositori di Galileo.* Vol. 3. *Cristoforo Scheiner.* Venice: Carlo Ferrari, 1919.

—. *Galileo Galilei e Suor Maria Celeste.* Florence: Giunti Barbèra, 1935.

—. *Galileo e lo Studio di Padova.* 3 vols. Padua: Antenore, 1966–68.

—. *Edizione Nazionale delle Opere di Galileo Galilei.* Reprint of the original edition, 1890–1909. Florence: Giunti Barbèra, 1968.

—. *Amici e corrispondenti di Galileo.* Reprint. Edited by P. Galluzzi. 3 vols. Florence: Salimbeni, 1983.

Feldhay, Rivka. *Galileo and the Church: Political Inquisition or Critical Dialogue?* Cambridge: Cambridge University Press, 1995.

————. "Recent Narratives on Galileo and the Church: On the Three Dogmas of the Counter-Reformation." In *Galileo in Context,* ed. Jürgen Renn, 219–38. Cambridge: Cambridge University Press, 2002.

Ferrone, Vincenzo. *Scienza, Natura, Religione.* Naples: Jovene, 1982.

Festa, Egidio. *Galileo: La lotta per la scienza.* Rome-Bari: Laterza, 2007.

Field, Judith V. "Cosmology in the Work of Kepler and Galileo." In Galluzzi, *Novità celesti e crisi del sapere,* 207–15.

Finocchiaro, Maurice A. *Galileo and the Art of Reasoning: Rhetorical Foundations of Logic and Scientific Method.* Boston: D. Reidel Publishing Company, 1980.

————. "The Methodological Background to Galileo's Trial." In *Reinterpreting Galileo,* ed. W. A. Wallace, 241–72. Washington DC: Catholic University of America Press, 1986.

————. *The Galileo Affair: A Documentary History.* Berkeley: University of California Press, 1989.

————. *Retrying Galileo, 1632–1992.* Berkeley: University of California Press, 2005.

————. *Defending Copernicus and Galileo: Critical Reasoning in the Two Affairs.* Boston Studies in the Philosophy of Science 280. New York: Springer-Verlag, 2010.

Firpo, Luigi. *Il Processo di Giordano Bruno.* Rome: Salerno Editore, 1993.

Frajese, Vittorio. *Il processo a Galileo Galilei: Il falso e la sua prova.* Brescia: Morcelliana, 2010.

Frova, Andrea, and Mariapiera Marenzana. *Thus Spoke Galileo: The Great Scientist's Ideas and Their Relevance to the Present Day.* Oxford: Oxford University Press, 2006.

Galilaeana: Journal of Galileo Studies. Edited by M. Buccianini and M. Camerota. Florence: Leo Olschki, 2004–.

Galilei, Galileo. *Opere.* 20 vols. Edited by Antonio Favaro. Florence: Barbèra, 1890–1909.

Galluzzi, P., ed. *Novità celesti e crisi del sapere: Atti del convegno internazionale di studi Galileiani.* Florence: Giunti Barbèra, 1984.

Garin, Eugenio. *L'educazione in Europa (1400–1600).* Bari: Laterza, 1957.

————. *Scienza e vita civile nel Rinascimento italiano.* Bari: Laterza, 1965.

————. "A proposito di Copernico." *Rivista di Storia della Filosofia* 26 (1971): 79–96.

————. "Alle origini della polemica copernicana." In *Studia Copernicana 6,* 31–42. Wroclaw: Ossolineum, 1973.

Geymonat, Ludovico. *Galileo Galilei.* Translated by S. Drake. New York: McGraw-Hill, 1965.

Gingerich, Owen. *The Book Nobody Read*. New York: Walker, 2004.

Grant, Edward. *A Source Book in Medieval Science*. Cambridge, MA: Harvard University Press, 1974.

———. "Cosmology." In *Science in the Middle Ages,* ed. David C. Lindberg, 265–302. Chicago: University of Chicago Press, 1978.

———. *Physical Science in the Middle Ages*. Cambridge, MA: Harvard University Press, 1979.

———. *In Defense of the Earth's Centrality and Immobility: Scholastic Reaction to Copernicanism in the Seventeenth Century*. Philadelphia: American Philosophical Society, 1984.

Heath, Thomas. *Aristarchus of Samos*. New York: Dover Publications, 1981.

Heilbron, J. L. *Galileo*. Oxford: Oxford University Press, 2010.

Hooykaas, Reijer. "Rheticus's Lost Treatise on Holy Scripture and the Motion of the Earth." *Journal for the History of Astronomy* 15 (1984): 77–80.

Jacobs, L. "Jewish Cosmology." In *Ancient Cosmologies,* ed. M. Loewe and C. Blacker, 69. London: George Allen & Unwin, 1975.

John Paul II. "Discourse on the One Hundredth Anniversary of the Birth of Albert Einstein." *Acta Apostolicae Sedis*. Vatican City State: Tipografia Poliglotta Vaticana, 1979. English translation in Bucciarelli, "Speech of His Holiness."

———. "Discourse to the Symposium, Galilean Studies Today, on the Occasion of the Commemoration of the 350th Anniversary of the Publication of the *Dialogue on the Two Chief World Systems*." *L'Osservatore Romano,* May 9–10, 1983, 1 and 3.

———. "Discourse to the City of Pisa." *L'Osservatore Romano,* September 24, 1989, 4.

———. "Discourse on the Occasion of the Plenary Session of the Pontifical Academy of Sciences and the Conclusion of the Work of the Study Commission on the Ptolemaic-Copernican Controversy." *Origins* 22 (November 12, 1992): 370–75.

Kepler, Johannes. *Gesammelte Werke*. Edited by Walter van Dyck, Max Caspar, and Franz Hammer Munich: C. H. Beck, 1937–93.

Koestler, Arthur. *The Sleepwalkers*. London: Hutchinson, 1959.

Koyré, Alexandre. *From the Closed World to the Infinite Universe*. Baltimore: Johns Hopkins University Press, 1957.

———. *Etudes d'histoire de la pensée scientifique*. Paris: Hermann, 1966.

———. *The Astronomical Revolution*. Translated by R. E. W. Maddison. London: Methuen, 1973.

———. *Galileo Studies*. Translated by J. Mephan. Hassocks: Harvester Press, 1978.

Kuhn, Thomas. *The Copernican Revolution.* Cambridge, MA: Harvard University Press, 1971.

Langford, Jerome J. *Galileo, Science, and the Church.* New York: Desclee Company, 1966.

Lattis, James M. *Between Copernicus and Galileo.* Chicago: University of Chicago Press, 1994.

Le Bachelet, Xavier Marie. "Bellarmin et G. Bruno." *Gregorianum* 4 (1923): 193–210.

Lerner, Michel-Pierre. "L'Hérésie' héliocentrique: Du supçon à la condamnation." In *Sciences et religions: De Copernic à Galilée,* 411–42. Rome: Collection de l'École Française de Rome, 1996.

———. "Pour une édition critique de la sentence et de l'abjuration de Galilée." *Revue des Sciences Philosophiques et Théologiques* 82, no. 4 (October 1998): 607–29.

———. "Vérité des philosophes et vérité des théologiens selon Tommaso Campanella o.p." *Freiburger Zeitschrift für Philosophie und Theologie* 48 (2001): 281–300.

———. "La reception de la condamnation de Galilée en France au XVII siècle." In Montesinos and Solís, *Largo campo di filosofare,* 513–47.

Loyola, Ignatius. *The Constitutions of the Society of Jesus.* St. Louis: Institute of Jesuit Sources, 1996.

Luther, Martin. *Tischreden.* In *D. Martin Luthers Werke.* Weimar: Böhlau, 2000–.

Maccarrone, Michele. *Mons. Paschini e la Roma eclesiastica: Atti del convegno di studio su Pio Paschini nel centenario della nascita, 1878–1978.* Pubblicazioni della Deputazione di Storia Patria del Friuli 10. Udine: Tipografia Poliglotta Vaticana, 1980.

MacColley, Grant. "Ch. Scheiner and the Decline of Neo-Aristotelianism." *Isis* 32 (1942): 63–69.

Maffei, Paolo. "Il sistema copernicano dopo Galileo e l'ultimo conflitto per la sua affermazione." *Giornale di Astronomia* 1 (1975): 5–12.

———. *Giuseppe Settele, il suo diario e la questione Galileiana.* Foligno: Edizioni dell'Arquata, 1987.

Maffeo, Sabino. *The Vatican Observatory: In the Service of Nine Popes.* Vatican City State: Vatican Observatory Publications, 2001.

Martinez, Rafael. "Il manoscritto ACDF, Index, Protocolli, vol. EE, f. 291r–v." *Acta Philosophica* 10, no. 2 (2001): 215–42.

Martini, Carlo M. "Gli esegeti al tempo di Galileo." In *Nel Quarto Centenario della Nascita di Galileo Galilei,* 115–124. Milan: Vita e Pensiero, 1966.

Mayaud, Pierre-Noël. "Une nouvelle affaire Galilée?" *Revue d'Histoire des Sciences* 45 (1992): 2–3.

————. "Les 'Fuit Congregatio Sancti Officii in . . . coram . . .' de 1611 à 1642: 32 ans de vie de la Congrégation du Saint Office." *Archivum Historiae Pontificiae* 30 (1992): 231–89.

————. *La condamnation des livres coperniciens et sa révocation.* Rome: Pontifical Gregorian University, 1997.

McMullin, Ernan, ed. *Galileo, Man of Science.* New York: Basic Books, 1967.

————. "Bruno and Copernicus." *Isis* 78 (1987): 55–74.

————. "Galileo on Science and Scripture." In *The Cambridge Companion to Galileo,* ed. Peter Machamer, 271–347. New York: Cambridge University Press, 1998.

————, ed. *The Church and Galileo.* Notre Dame, IN: University of Notre Dame Press, 2005.

Mercati, Angelo. *Il sommario del processo di Giordano Bruno con appendice di documenti sull'eresia e l'inquisizione a Modena nel secolo XVI.* Vatican City State: Vatican Apostolic Library, 1942.

Mereu, Italo. *Storia dell'Intolleranza in Europa.* Milan: Bompiani, 1990.

Migne, J.-P., ed. *Patrologiae Cursus Completus. Series Prima Latina.* Paris: MPL, 1865.

Montesinos, J., and C. Solís, eds. *Largo campo di filosofare: Eurosymposium Galileo 2001.* Orotava: Fundación Canaria Orotava de Historia de la Ciencia, 2001.

Morpurgo-Tagliabue, Guido. *I processi di Galileo e l'epistemologia.* Milan: Edizione di Comunità, 1963.

Moss, Jean D. *Novelties in the Heavens: Rhetoric and Science in the Copernican Controversy.* Chicago: University of Chicago Press, 1993.

Norlind, Wilhelm. "Copernicus and Luther. A Critical Study." *Isis* 44 (1954): 273–76.

————. "A Hitherto Unpublished Letter from Tycho Brahe to Christopher Clavius." *The Observatory* 74 (1954): 20–23.

————. "Tycho Brahe et ses rapports avec l'Italie." *Scientia* (February 1955): 1–15.

————. *Tycho Brahe: A Biography.* Lund: C. W. K. Gleerup, 1970.

Oreggi, A. *De Deo uno tractatus primus.* Rome: 1629.

Oresme, Nicole, *Le livre du ciel et du monde.* Translated by Albert D. Menut. Madison: University of Wisconsin Press, 1968.

Pagano, Sergio M., ed. *I Documenti vaticani del processo di Galileo Galilei (1611–1741).* Vatican City State: Archivio Segreto Vaticano, 2009.

Pantin, Isabelle. *Sidereus Nuncius Le Messager Celeste.* Paris: Les Belles Lettres, 1992.

————. *Kepler: Discussion avec le Messager Celeste.* Paris: Les Belles Lettres, 1993.

Paschini, Pio. *Vita e Opere di Galileo Galilei*. Rome: Herder, 1965.

Pastor, Ludwig, Freiherr von. *History of the Popes from the Close of the Middle Ages*. Vol. 28. Translated by E. Graf. London: Routledge & Kegan Paul, 1938.

Pedersen, Olaf. *Galileo and the Council of Trent*. Vatican City State: Vatican Observatory Publications, 1991.

———. *The Book of Nature*. Vatican City State: Vatican Observatory Publications, 1992.

Pedersen, Olaf, and Mogens Pihl. *Early Physics and Astronomy*. Revised ed. Cambridge: Cambridge University Press, 1993.

Pesce, Mauro. *L'ermeneutica biblica di Galileo e le due strade della teologia cristiana*. Rome: Edizioni di Storia e Letteratura, 2005.

Poppi, Antonino. *Cremonini, Galilei e gli inquisitori del Santo a Padova*. Padua: Centro Studi Antoniani, 1993.

Poupard, Paul, ed. *Toward a Resolution of 350 Years of Debate—1633–1983*. Pittsburgh: Duquesne University Press, 1987.

Prosperi, Adriano. *Tribunali della coscienza*. Turin: Einaudi, 1996.

Renn, Jürgen, ed. *Galileo in Context*. Cambridge: Cambridge University Press, 2002.

Redondi, Pietro. *Galileo: Heretic*. Translated by R. Rosenthal. Princeton: Princeton University Press, 1983.

Reston, James, Jr. *Galileo: A Life*. New York: Harper Collins, 1999.

Ronan, A. "The Origins of the Reflecting Telescope." *Journal of the British Astronomical Association* 101 (1991): 335–42.

Ronchi, Vasco. *Il cannocchiale di Galileo e la scienza del seicento*. 2nd ed. Turin: Einaudi–Boringhieri, 1958.

———. "Galilée et l'astronomie." In *Galilée: Aspects de sa vie et de son oeuvre*, ed. Suzanne Delorme, 153–72. Paris: Presses Universitaires de France, 1968.

———. *Storia della Luce: Da Euclide a Einstesin*. Bari: Laterza, 1983.

Rosen, Edward. *Three Copernican Treatises*. New York: Dover Publications, 1971.

———. "Was Copernicus's *Revolutions* Approved by the Pope?" *Journal of the History of Ideas* 36 (1975): 531–42.

———. *Copernicus and the Scientific Revolution*. Malabar, FL: Krieger Publishing, 1984.

Rossi, Paolo. *Galileo Galilei*. Rome: Istituto Poligrafico e Zecca dello Stato, 1997.

Rowland, Wade. *Galileo's Mistake: A New Look at the Epic Confrontation between Galileo and the Church*. New York: Arcade Publishing, 2003.

Santillana, Giorgio de. *The Crime of Galileo*. Chicago: University of Chicago Press, 1955.

Segre, Michael. *In the Wake of Galileo.* New Brunswick: Rutgers University Press, 1991.

———. "Light on the Galileo Case?" *Isis* 88 (1997): 484–504.

Settle, Thomas B. "Experimental Sense in Galileo's Early Works and Its Likely Sources." In Montesinos and Solís, *Largo campo di filosofare,* 831–49.

Sharratt, Michael. *Galileo: Decisive Innovator.* Cambridge: Cambridge University Press, 1996.

Shea, William R. *Galileo's Intellectual Revolution: Middle Period, 1610–1632.* New York: Science History Publications, 1972.

———. "Melchior Inchofer's *Tractatus Syllepticus*: A Consultor of the Holy Office Answers Galileo." In Galluzzi, *Novità celesti e crisi del sapere,* 283–92.

———. "Galileo the Copernican." In Montesinos and Solís, *Largo campo di filosofare,* 41–59.

Shea, William R., and Mariano Artigas. *Galileo in Rome: The Rise and Fall of a Troublesome Genius.* Oxford: Oxford University Press, 2003.

———. *Galileo Observed: Science and the Politics of Belief.* Sagamore Beach, MA: Science History Publications, 2006.

Simoncelli, Paolo. *Storia di una censura: "Vita di Galileo" e Concilio Vaticano II.* Milan: Franco Angeli, 1992.

Sobel, Dava. *Galileo's Daughter.* New York: Penguin Books, 1999.

Soccorsi, Filippo. *Il Processo di Galileo.* Rome: Edizioni La Civiltà Cattolica, 1963.

Speller, Jules. *Galileo's Inquisition Trial Revisited.* Frankfurt am Main: Lang, 2008.

Thomas Aquinas. *Summa Theologica.* Amer. ed. 3 vols. New York: Benziger Brothers, 1947.

———. *Commentarium de Coelo et Mundo.* Turin: Marietti, 1953.

Thoren, Victor E. *The Lord of Uraniborg: A Biography of Tycho Brahe.* New York: Cambridge University Press, 1990.

Torrini, Maurizio. "Galileo e la repubblica degli scienziati." In Montesinos and Solís, *Largo campo di filosofare,* 783–94.

Vernet, Juan. "Copernicus in Spain." In *Colloquia Copernicana I: Studia Copernicana 5,* 271–91. Wroclaw, 1972.

Viganò, Mario. *Il Mancato Dialogo fra Galileo e i Teologi.* Rome: Edizioni La Civiltà Cattolica, 1969.

Wallace, William A. *Galileo's Early Notebooks: The Physical Questions.* Notre Dame, IN: University of Notre Dame Press, 1977.

———. "Galileo Galilei and the Doctores Parisienses." In *New Perspectives on Galileo,* ed. R. E. Butts and J. C. Pitt, 87–138. Dordrecht: D. Reidel Publishing Company, 1978.

————. "The Philosophical Setting of Medieval Science." In *Science in the Middle Ages,* ed. David C. Lindberg, 91–119. Chicago: University of Chicago Press, 1978.

————. *Prelude to Galileo: Essays on Medieval and Sixteenth-Century Sources of Galileo's Thought.* Dordrecht: D. Reidel Publishing Company, 1981.

————. *Galileo and His Sources.* Princeton: Princeton University Press, 1984.

————. "Galileo's Early Arguments for Geocentrism and His Later Rejection of Them." In Galluzzi, *Novità celesti e crisi del sapere,* 31–40.

————, ed. *Reinterpreting Galileo.* Washington, DC: Catholic University of America Press, 1986.

Westfall, Richard S. *Essays on the Trial of Galileo.* Studi Galileiani 1.5. Vatican City State: Vatican Observatory Publications, 1989.

Westman, Robert. "The Reception of Galileo's *Dialogue*: A Partial World Census of Extant Copies." In Galluzzi, *Novità celesti e crisi del sapere,* 329–37.

Wisan, Winifred L. "On the Chronology of Galileo's Writings." *Annali dell'Istituto e Museo della Scienza di Firenze* 9, no. 2 (1984): 85–88.

Wootton, David. *Galileo: Watcher of the Skies.* New Haven: Yale University Press, 2010.

Yates, Frances A. *Giordano Bruno and the Hermetic Tradition.* Chicago: University of Chicago Press, 1969.

Zycinski, Jozef. *The Idea of Unification in Galileo's Epistemology.* Studi Galileiani 1.4. Vatican City State: Vatican Observatory Publications, 1988.

An internationally known Galileo scholar,
Annibale Fantoli is adjunct professor of philosophy at the
University of Victoria. An English edition of his biography
Galileo: For Copernicanism and for the Church *was published*
in 1994 by the Vatican Observatory and distributed by the
University of Notre Dame Press. It has since reached
a third edition and been translated into several languages.